T0065678

evolution
from SPACE

A THEORY OF COSMIC CREATIONISM

Sir Fred Hoyle
N.C. Wickramasinghe

A TOUCHSTONE BOOK
Published by Simon & Schuster, Inc.
NEW YORK

First Touchstone Edition, 1984
Published by Simon & Schuster, Inc.
Simon & Schuster Building
Rockefeller Center
1230 Avenue of the Americas
New York, New York 10020
Published by arrangement with J. M. Dent & Sons Ltd.
TOUCHSTONE and colophon are registered trademarks
of Simon & Schuster, Inc.

Manufactured in the United States of America

10 9 8 7 6 5 4 3 2 1
10 9 8 7 6 5 4 3 2 1 Pbk.

Library of Congress Cataloging in Publication Data

Hoyle, Fred, Sir.
  Evolution from space.

  Reprint. Originally published: London:
Dent, 1981.
  Includes index.
  1. Evolution. 2. Life-Origin.
3. Cosmogony. I. Wickramasinghe, N. C.
(Nalin Chandra), 1939-   . II. Title.
QH371.H68   1982   577   82-5622
                         AACR2

ISBN 978-0-671-49263-2

Contents

List of figures

Acknowledgments

The authors and publishers would like to thank the following for permission to quote extracts or reproduce diagrams from their publications, as indicated in the text:

W. A. Benjamin Inc.: J. D. Watson, *Molecular Biology of the Gene*, 1977.

Academic Press: ed. D. J. Kushner, *Microbial Life in Extreme Environments*, 1978.

BBC Publications and Collins: David Attenborough, *Life on Earth*, 1979.

Edward Arnold: G. M. Bennison and A. E. Wright, *The Geological History of the British Isles*, 1969.

Work Publ. Inc.: A. L. Lehninger, *Biochemistry*, 1978.

Encyclopedia Britannica: diagram from article entitled 'Insects'.

Clarendon Press: B. John and K. R. Lewis, *Chromosome Hierarchy*, Oxford, 1975.

Springer Verlag, New York: S. Ohno, *Evolution*, 1970.

W. H. Freeman & Co.: T. Dobzhansky, S. J. Ayala, G. L. Stebbins and J. W. Valentine, *Evolution*, 1977.

E. P. Dutton & Co. Inc.: Loren Eiseley, *Darwin and the Mysterious Mr X*, 1979.

The Chairman of the Sri Lanka Broadcasting Corporation for permission to reproduce an extract from a broadcast interview.

Note on technical terms

When either very large or very small numbers have to be written it is convenient to use powers of ten in the compact form illustrated by the following table:

$$10^{10} = 10,000,000,000$$
$$10^{9} = 1,000,000,000$$
$$10^{8} = 100,000,000$$
$$10^{7} = 10,000,000$$
$$10^{6} = 1,000,000$$
$$10^{5} = 100,000$$
$$10^{4} = 10,000$$
$$10^{3} = 1,000$$
$$10^{2} = 100$$
$$10^{1} = 10$$
$$10^{0} = 1$$
$$10^{-1} = 0.1$$
$$10^{-2} = 0.01$$
$$10^{-3} = 0.001$$
$$10^{-4} = 0.0001$$
$$10^{-5} = 0.00001$$
$$10^{-6} = 0.000001$$
$$10^{-7} = 0.0000001$$
$$10^{-8} = 0.00000001$$
$$10^{-9} = 0.000000001$$
$$10^{-10} = 0.0000000001$$

Besides expressing small numbers with the decimal point we can also think in terms of fractions. If a rich man's fortune is divided into a thousand equal portions, and if we happen to inherit one such portion, then we receive a fraction equal to one part in a thousand, 10^{-3}, of the original fortune.

There is no difficulty in extending the notation to numbers that are not exact powers of 10. Thus the number 280,000 (two hundred and eighty thousand) would amount to the following: $(280) \times (1000) = (2.8 \times 10^{2}) \times (10^{3}) = 2.8 \times 10^{5}$. This illustrates the rule that powers of 10 add in multiplication. On the other hand, powers of 10 subtract in division, as in the following:
$(800) \div (4000) = (8 \times 10^{2}) \div (4 \times 10^{3}) = (8 \div 4) \times 10^{2-3} = 2 \times 10^{-1}$.

A number $10^{40,000}$ appears frequently in this book. Forty thousand zeros would be needed to write this number in full, which at sixty zeros to a line would occupy rather more than sixteen pages of zeros. This example makes the advantage of the compact power notation very obvious.

As to measurement, the following equivalents are relevant:

1 micron or micrometre (μm) = 1 millionth of a metre = 10,000 Ångströms

1 Ångström = 10^{-10} metre (1 ten billionth of a metre)

The Ångström (A) is used for the measurement of wavelengths of light and intramolecular distances.

Introduction

It was widely held until the mid-seventeenth century that the Earth was located at the centre of the whole universe. Copernicus (1473-1543) had thought otherwise, but his views made no great impact for a hundred years, and they might have taken even longer to become widely accepted but for the rise of the Protestant movement in N.W. Europe. The issue was as much a matter of religion and sociology as it was of science. Today of course we know there is no specially favoured position in the universe. If one wishes to think of a 'centre', as indeed it is convenient to do in tackling some problems, any place one cares to choose for the 'centre' will serve just as well as any other place. Our modern universe is vast in its scale compared to the medieval world—to be precise it is 100,000,000,000 times greater in its spatial range and more than a million times greater in its span of time.

While our escape in solar-system astronomy from the blinkered medieval world is a reason for a certain measure of self-congratulation, we must be somewhat sparing in the adulations we heap upon ourselves, since almost exactly the same pre-Copernican errors are being made nowadays in the frontier area of astronomy, chemistry and biology. There is the same passionate desire for the Earth to be the centre of life, the focal point of chemical evolution based on the carbon atom, that there used to be for geometry and astronomy. This is in spite of a great deal of evidence to the contrary, in spite of free schooling for everybody, in spite of free libraries, of a plethora of universities and colleges and in spite of the huge funds made available by governments for scientific research.

Once again the issue is dominated more by sociology and religion than by science. More precisely, dominated by antireligion. As man passed from agriculture to an urban-machine culture (with the industrial revolution of the nineteenth century) he became more self-reliant, more conscious of being in control of his own destiny. Except when whole populations were required to present themselves for mass slaughter in times of war, it was found economically advantageous to write religion out of everyday life. Churches, dotted everywhere over the land, emptied until they were scarcely one-tenth full of a Sunday. They became quaint monuments to a bygone age.

The machines provided the answers to everyday problems (except for those which afflict the old and infirm, whose voices grow too enfeebled to be heard). Not all problems, however, were of an everyday urgency. There was the problem of the origin of man himself. Already in the nineteenth century this problem was seen to be connected with the origin of other animals, and indeed with the origin of life as a whole. Since there was no reason of logic, observation or experiment that required life to be terrestrial rather than cosmic in its beginning, an unbiased investigator would have chosen to consider both possibilities with an open mind as to the eventual outcome. Instead almost everyone plunged helter-skelter along the terrestrial route, and the few who paused to take a look outside the Earth became regarded as figures of fun, when in truth the situation was just the opposite way around.

Things have become steadily more pre-Copernican as time has gone along. It is understandable that the biologists of the nineteenth and early twentieth centuries did not appreciate the extreme chemical complexity of living systems. Without doing serious violence to the evidence it was then possible to think in terms of processes leading to the origin of life that were not much more subtle than those which take place naturally in the Earth's crust, as for instance the geochemical processes which led to the formation of metallic ores. Thus it was thought that life might have arisen by spontaneous physical and chemical effects, and that similar processes could also have occurred on planets other than the Earth, everywhere throughout the universe. If such a view could have been maintained, the Earth would not have been elevated to the life-centre of the universe, and the position would not have been objectionably pre-Copernican in its logic. This was the situation in 1924 at the time of the work of A.I. Oparin, work that was widely acclaimed as putting the final nail in the coffin of the older religions. All of life (and death) could be seen, it was claimed, to spring from natural causes.

With the development of microbiology in the second half of the twentieth century it became overwhelmingly clear that the truth is quite otherwise. Biochemical systems are exceedingly complex, so much so that the chance of their being formed through random shufflings of simple organic molecules is exceedingly minute, to a point indeed where it is insensibly different from zero.

The new evidence pointed clearly and decisively to a cosmic origin of life, which is to say to the alternative possibility that had been available from the beginning. Perhaps because it was seen that a switch to this alternative might start filling the churches again, the response to the new evidence was to follow a far more perverse logic than before. It was agreed that the chance of life arising on the Earth was utterly minuscule, but instead of accepting the obvious corollary, it was argued that the minute chance proves that life can exist nowhere else but on the Earth. So one arrives at the pre-Copernican view widely held today. This book and our two previous ones attempt to protest against this view.[1]

Those who continued to struggle to church always had one solid point going for them. How could such a highly organized structure as a human, or a mouse, horse, or flower for that matter, have emerged from the largely disordered situation that existed four and a half billion years ago at the time of formation of the Earth? The problem with this question is to quantify it. Nobody can deny that information would be needed to make a flower or a human, but exactly how much information? Precisely because the amount would be large, going far outside normal experience, our traditional mathematical systems are not well-designed to cope with this problem. For example, the well-known concept of 'entropy' arose from the need to understand the operation of steam engines, which are surely many orders simpler than a mouse. It need occasion no surprise therefore that classical methods fail to answer the question.

By drawing on ideas from computer technology, however, we can make a little progress. The genetic instructions (genome) required to specify a particular life-form can be thought of as a message of certain length, the length being the minimum that is sufficient to convey the required message. Then one can work out without undue difficulty the total number of messages of the same length (whether meaningful or not) that could be constructed. This number can then be regarded as the information content of the original message. In effect, information content is determined by the number of ways in which the original message might have been wrongly specified. Similarly, one can think of

[1] *Lifecloud*, J.M. Dent, London, 1978; *Diseases from Space*, J.M. Dent, London, 1979.

3

the information content of a telephone number. Suppose it is urgent to contact an acquaintance whose number you have forgotten. How many times do you have to dial wrong numbers to find your acquaintance? This would represent the information content of the correct telephone number, if you had happened to remember it.

Interpreted in such a way, the genomes of mice, humans, flowers ... are enormous, fantastic, quite out of all nonbiological experience. This issue will arise repeatedly, particularly in chapters 2, 7, and 9.

In contrast, the primitive Earth was simple enough for its information content to be determined by classical methods, the methods invented to understand steam engines. There is a huge gulf in this respect between inorganic systems and even the most humble forms of life, which means that much explaining remains to be done. Evolution is the preferred mode of explanation, specifically Darwinian evolution, according to which life-forms change by 'natural selection'.

To understand the concept of evolution let us begin with the observation that the individuals of any species, born in a particular time interval, are not exactly identical to each other. For thousands of years arable farmers and stockmen have used this fact to 'select' plants and animals according to some desired end. This is done by breeding in each generation only from those individuals that best meet the chosen criterion of selection, which might be a faster horse, a stronger ox, or a more nutrient variety of wheat.

Sometime before 5000 BC, probably in upland villages of the Near East, it was by selecting wild grasses for improved nutrient value and for an improved yield of their seeds that agriculture was born. Everything known about the cultural patterns of semi-nomadic peoples suggests that this crucial step was taken by women, at times when their menfolk were occupied on inconsequential hunting expeditions. Nowadays it is rather common for girls in mixed-sex schools to experience social pressure from the boys not to study the sciences. Any girl with an interest in science should remember with scorn, if subject to such pressure, that it was almost surely the discovery of agriculture by women which permitted society to become civilized.

This artificial process of 'selection' is made possible by variations between individuals of the same species, variations which arise from small differences in the genetic messages which have been inherited from slightly different parents in the preceding generation. Practical experience in both plant-breeding and animal-breeding shows that selection for a particular characteristic cannot continue indefinitely. Once the form of gene, or of the set of genes, best suited to the characteristics in question

4

has been fully selected, improvement stops. 'Artificial selection' consists therefore in giving special emphasis to a particular form of genetic message that is already present in a species, without any reference to how the particular form of message got there in the first place.

It became evident to naturalists in the nineteenth century, of whom Edward Blyth appears to have been the first to formulate the basic issues explicitly, that a similar process must take place in the wild, thus making for a 'natural process of selection'. It was seen that if natural selection were to operate like artificial selection, only by emphasizing already-present characteristics, not much in the way of evolution could be achieved. Particular characteristics might be emphasized to a certain degree, and then no more. It was therefore necessary, if evolution was to be really important in the wild state, that genes should be able to change spontaneously in such a way that the genes of the offspring were not entirely determined by the genes of the parents.

Blyth had the idea, followed later by Charles Darwin, that variations in the offspring might be caused by the stress the parents had experienced in the wild state—the greater the stress to the parents, the greater the variations which appeared in the offspring. This was one of the few ideas of Blyth that turned out to be incorrect. It undoubtedly had its origin in the evolutionary theory of J.B. de Lamarck, according to whom the genetic message transmitted to the offspring was significantly affected by the mode of life of the parents. Many experiments have shown Larmarck's theory to be wrong. While it is true that the genes of the offspring may differ from the genes of the parents, the cause is much simpler and more straightforward than Lamarck's suggestion. Just as even the most accurate electronic copier occasionally garbles a message, so a gene is occasionally garbled as it is copied from parent to offspring.

Modern experiments have shown the garbling rate to be extremely slow, however, far too slow to provide variations that would be sufficient for natural selection to be much different from artificial selection. It is true that evolutionary texts from Darwin onwards contain examples of animals which appear to adapt themselves rapidly to a changing environment. The examples are often concerned with the colours of birds and insects which alter so as to keep a match with some aspect of the physical environment. There are also examples in which bacteria appear to adapt very quickly so as to become resistant to the particular drugs used in hospitals. We doubt, however, that anything more is involved in these cases than the selection of already existing genes. They are simply unusual cases of artificial selection.

Besides which, the sceptical reader will wonder how it would ever be possible to generate information out of a garbling process. Garbling overwhelmingly degenerates information. But not indeed in every case. There is a very small, not strictly zero, probability that the garbling of a message will turn out to improve it. So if one has a means for noticing and grabbing hold of the improved copy in the rare case in which it happens, the information content of the message can actually be improved. This is the logic of the Darwinian theory of evolution.

The logic turns on the ability to recognize the rare improved copy. In the case of a normal linguistic message the ability would need to reside in the brain of the person who noticed the improved copy. From which we see that no new information would really be generated, since the information of what constituted an improved copy had to be already there in a human brain before the garbling of the message occurred. What is it, one can wonder, that corresponds in the Darwinian theory to the human brain for the case of a linguistic message? What is it that seizes hold of the improved version of a plant or animal?

The answer given to this crucial question is that it is survival in the environment which makes the grab. But then how could the physical environment, containing what seems to be only a low level of information, succeed in building biological systems with exceedingly high levels of information? The survival of a species, with respect to both direct physical hazards and to competition from other species, is supposed to provide the answer to this further question. Whether it does so or not has always been a matter of debate. Quite certainly there is nothing in survival as such which guarantees the accumulation of information. Rocks from the Archean period of geology have survived for much longer than any multicelled life-form, and yet they contain comparatively little information.

Geological systems, however, are not in competition with each other in the same way as biological systems. Thus one can argue that it is competition which sets the informational standard for survival, rather than simple physical hazards. Species evolve by working against each other. To get ahead of the competition we must 'improve', which is to say we must increase our informational standard. So too must all others who survive. Hence the outcome, among the survivors, is for informational standards to increase all round. This is the essence of the Darwinian theory of evolution.

But without any concession to logic, the argument can be stated inversely. If the standard of one's competitors declines, one can afford to decline oneself and still survive. Because the same is true for every species, survival is also possible in a generally worsening situation. Thus by assuming implicitly that the competition does not decline, the Darwinian theory really begs the question. This inversion of the argument is by no means a quibble. We saw above that the variations on which natural selection operates arise from the miscopying of genetic messages. Miscopying commonly loses information and gains it only rarely. The variations on which natural selection operates are therefore strongly biased towards decline.

Nothing much more can be achieved by a general argument in words. Mathematics is needed to make further progress. Although the relevant mathematical results only became known in recent years,[2] they are not particularly difficult to derive, and it is something of a mystery to understand why they were not obtained long ago. Possibly the reason was that the results are not particularly helpful to the Darwinian theory.

The outcome depends crucially on the number of individuals which constitute the breeding group of a species or subspecies. If the number is large enough, as it may be for some bacteria and plants and some insects, the trend is Darwinian, towards improvement. Otherwise the trend is towards decline. This makes for difficulty in understanding the evolution of reptiles, birds and mammals.

But in any case the difficulty was there already in a concealed form. It is all very well to argue that the informational standard is maintained by one's competitors, but how did one's competitors acquire their information in the first place? From their forebears, it would be argued, pushing the problem back a generation. Indeed, in each generation one has to argue back to the preceding generation, with the hope of arriving eventually at a much lower informational standard, which might conceivably have come directly from the physical environment.

Readers with experience of practical 'business' will have realized that everything which has been said about biological evolution could be taken over almost *mutatis mutandis* to the evolution of industry. Modern industrial corporations have higher information standards than their predecessors, due, it can be seen, to a continuing competition between rival companies. Going backwards in time, decade by decade, things become simpler and simpler, until only twenty decades ago one is back in the era of cottage industries, which arose from the relation of individual humans to their environment. Thus the apparently complex organizational structure of modern industry can be explained, starting from a simple beginning, by an evolutionary process analogous to the

---

[2] J.L. King and T.H. Jukes, *Science,* 164, 1969, 788; J.F. Crow and M. Kimura, *An Introduction to Population Genetics Theory,* Harper and Row, New York, 1970; M. Kimura and T. Ohta, *Nature,* 229, 1971, 467.

Darwinian theory. (The analogy may not be fortuitous. We suspect that it was experience in a growing industrial society which really suggested the biological theory.)

The problem for biology is to reach a simple beginning. Going back in time to the age of the oldest rocks, more than eighty per cent of the age of the Earth itself, fossil residues of ancient life-forms discovered in the rocks do not reveal a simple beginning. Although we may care to think of fossil bacteria and fossil algae and microfungi as being simple compared to a dog or horse, the information standard remains enormously high. Most of the biochemical complexity of life was present already at the time the oldest surface rocks of the Earth were formed. Thus we have no clue, even from evidence which penetrates very far back in time, as to how the information standard of life was set up in the first place, and so the evolutionary theory lacks a proper foundation.

As soon as we turn from a terrestrially limited theory to a cosmic point of view all the difficulties mentioned above are either overcome or are mitigated in some degree. The Earth becomes a part of a vastly wider system, the universe itself, and it is the wider system which supplies the informational standard. The 'cottage industries' were not located on the Earth at all. Life had evolved already to a high information standard long before the Earth was born. We received life with the fundamental biochemical problems already solved. This is the point of view we shall seek to develop in the following chapters.

1
Blood

Scientific method is actually an illusory concept, but if one reads what passes for scientific method it goes something like this:

Stage 1   Observe and collate what happens in the world.

Stage 2   Using the information from Stage 1, form theories as to what may be true in the world.

Stage 3   Design experiments to test whether the theories of Stage 2 are really true or not.

The implicit assumption here is that Nature is a passive bystander, waiting there to be experimented on. Sometimes this may seem to be true. The earnest schoolboy who carries through an experiment to find out how much work makes so much heat gets an easy-tempered response. On other occasions, however, Nature answers with a lion-like roar, as if a teacher had at last lost patience with a dull-witted pupil. Such tempestuous replies make sport of scientific method.

In 1881 A.A. Michelson and E.W. Morley carried out an experiment the background of which was the following. Sound waves are possible only if there is air or some other material for them to pass through. One can't shout in a vacuum. When in 1864 James Clerk Maxwell established that light also travels as a wave it was felt that, just like sound, there must be some physical medium through which it moved. It was understood from the beginning that the medium could not be gas. Nobody knew what the medium was, so they gave it the Greek name *aether*. To put the situation briefly, whereas sound consists of a rhythmic shaking of the molecules of a gas, light was supposed to consist of a rhythmic shaking of *aether.*

Michelson and Morley designed their experiment to measure the speed of the Earth's motion through the *aether.* The concept was of the Earth voyaging through *aether* like a ship through the sea. From the manner in which the experiment was set up, the expected answer was the rotation speed of the Earth (as it happens to be at the geographical latitude of Pasadena in S. California, which was where the experiment was carried out).

If the answer had been twice what was expected, or a half, or some other multiple or fraction, physicists would have been seriously worried, but not nearly as worried as they were by the actual answer. The motion of Ship Earth through Ocean *Aether* turned out to be nil. And it was still nil six months later when the Earth's motion around the Sun had reversed in direction. No matter how the ship moved, the ocean always moved with it, so that the ship was perpetually in still water.

This set a terrible cat among the scientific pigeons, since what was true for the Earth would have to be true for every other body in the universe, and how could *aether* contrive to be stationary for many bodies all with different motions? It did not seem hard to convince oneself that such an idea would lead to hopeless contradictions, and the disposition therefore was to seek a loophole in the experiment. It was not for 23 years, until a lecture in 1904 given in St Louis, Missouri, that at last the French mathematician Henri Poincaré gave unequivocal acceptance to the situation. Then in 1905 Albert Einstein faced up to all the contradictions that were supposed to follow from the experiment. By painstakingly following through the logic in fine detail Einstein showed that, while the experiment certainly led to strange conclusions, it did not lead to any self-contradiction. (Einstein is often credited with having propounded the energy equivalent $E = mc^2$. This equivalence had more to do with H.A. Lorentz and M. Planck, however, and was first given fully by H. Minkowski. This is an example of the distortion of historical fact by media publicity. Einstein's great contribution lay in clearing up the confusion described in the text above.) So far from Nature giving a simple yes-or-no answer to the original question, Nature had delivered a bomb-blast, and by 1905 it had taken almost a quarter of a century to sort out the wreckage.

Nature may offer observations which have devastating implications and which yet need no experimental equipment at all. A straight stick standing in a pond appears to be bent at its point of emergence from the water. This observation, which Ernest Rutherford once said had started him on his career as a physicist, has been available to man since early prehistoric days. The similarities and differences between plants and animals, and indeed many aspects of the phenomenon of life, have also been available for observation since the earliest times. Whereas societies in the past have not shown much interest in, and have certainly not flown into passions over, an apparently-bent stick standing in a pond, there have always been strong feelings over man's place in the living world. The religions of the past, some of which we still have with us today, were founded in speculations on such matters.

It was a major triumph of the scientific naturalists of the first half of the nineteenth century, of whom Charles Darwin is much the best known, that they managed to replace the earlier wild speculations by an evolutionary theory with a rational form to it. The triumph of the scientific naturalists, and in particular of the theory of mutation and natural selection, came as an enormous relief to those many people who had found the concept of a universe based on trivialities quite painful to contemplate. This view is still held today, both among biologists and among those of a 'rationalist' turn of mind, which probably includes most workers in the physical sciences also. There is a widespread apprehension that any weakening from a Darwinian position would open the flood-gates to new waves of irrationalism, and that the former victory would then be turned to defeat. The real defeat, however, would come from maintaining a wrong position (if it is wrong) for sociological reasons. Victory in science comes only from moving in the direction in which Nature points, no matter how unwelcome from a human point of view that direction may seem. This is the true essence of 'scientific method', not the trite precepts set out at the beginning of this chapter.

Lord George Murray is said to have remarked before the battle of Culloden that 'there never could be more improper ground for Highlanders to fight on', and one might think there could be a no more improper scientific battleground for two astronomers to fight on than the Darwinian theory of natural selection, a theory supported overwhelmingly by the whole scientific world. But this is not the way we ourselves read Nature's signposts, and the ground is therefore not of our own choosing.

In this chapter we discuss three such signposts, three curious cases, in which various life-forms appear to possess properties that they have no right according to Darwinian theory to possess. We do this first for flies, secondly for peas and beans, and thirdly for bacteria.

Darwinism makes few precise statements, and for this reason it is a hard theory to nail. Faced with the evidence we discuss in this chapter, it immediately takes refuge in an indefinite situation, by arguing that environmental conditions might have been different from the way they are today, and different from the way we think they were in the past. If the only difficulties were the ones considered in this chapter, ways to evade them might conceivably be found along these lines. But these first difficulties are only a small beginning. In the next chapter we shall bring heavier guns to bear. Our main assault will come, however, in chapters 6, 7 and 8.

The very different form of evolution that we believe the evidence points decisively towards is briefly explained in chapter 2, with a more extended discussion of the technical details held over to chapters 3 to 5, and with a discussion of the deeper implications reserved until chapter 9.

In these days of hastily written scientific papers which aim at comprehensibility only to specialist readers, it is a delight to read 'Reactions of *Drosophila* to 2537 Å Radiation' by Frank E. Lutz and E.N. Grisewood. Ultraviolet radiation with a wavelength of 2537 Ångströms was produced by a mercury arc lamp.[1] Humans do not see ultraviolet radiation of this wavelength, and if it were to enter our eyes much damage would be caused thereby. Fortunately, this does not happen under natural conditions because the ozone layer in the Earth's upper atmosphere removes this particular region of the ultraviolet from the spectrum of the Sun. Thus apart from man-made generators of 2537 Å, such as the mercury arc lamp, there is no radiation of this wavelength at the Earth's surface, and therefore no natural environment to have caused insects to develop vision at 2537 Å by biological selection. Yet Lutz and Grisewood found *Drosophila melanogaster* to be visually responsive to 2537 Å with a sensitivity comparable to their response to ordinary light.

The possibility of a fluorescent process, whereby 2537 Å served to generate ordinary light, which was then recognized by the flies, had next to be considered. Lutz and Grisewood showed not only that this did not happen, but that the outer eyes of bees and larger flies (*Drosophila* being too small for this further experiment) are transparent to 2537 Å, which must therefore reach the inner eyes of these creatures.

Lutz and Grisewood concluded:

> In other words (the indications are that) normal *Drosophila melanogaster* can see 2537 Å radiation directly and was not reacting to

[1] Along with radiation of other wavelengths which Lutz and Grisewood removed with the aid of a spectrograph so as to be left only with the 2537 Å component in their experiment: *American Museum Novitates,* no. 706, March 1934.

an indirect effect of it. If this be true, it has some interesting bearings on biological speculation in view of the fact that light of so short a wavelength does not and probably never did occur in the environment of insects.

The year 1934 when this work was published is now a long time ago, and one would suppose any uncertainties there might have been in it would long ago have been removed by later experiments. If so, we have not been able to trace them in the literature. Nor does the classic text *The Principles of Insect Physiology* by V.B. Wigglesworth give any reference to later investigations of this remarkable phenomenon.[2] Indeed, Wigglesworth comments:

> The honey-bee will respond to light with a wavelength of 2970 Å, at the lower limit of the solar spectrum; and *Drosophila* will undoubtedly respond even to 2537 Å.

The old saying 'You can't get blood out of a stone' appears at first sight really to be true, but probably it isn't. Likely enough, some stones can be found that contain all the chemical elements present in blood. Possibly some day a chemist will amuse him or herself by putting them together into hemoglobin. Likewise one might think that 'You can't get blood out of a pea or a bean' would be true, but it certainly is not.

In 1938, J. Pietz obtained a red pigment from *Vicia Faba,* a member of the pea family. In the following year H. Kubo obtained the same red pigment from a wide variety of leguminous plants. Kubo also found certain spectroscopic 'fingerprints' which characterized the pigment as a hemoprotein, and he believed it to be hemoglobin. The second world war then intervened, and it was not until 1945 that Kubo's opinion was confirmed by D. Keilin and Y.L. Wang.[3]

It will be as well to insert a short discussion on the structure of hemoglobin and of one or two related biomolecules. The essential feature of these molecules is the so-called heme group of atoms (about one hundred of them) of which the crucial component is an iron atom bonded to four nitrogen atoms. A quite similar group of about one hundred atoms, but with magnesium replacing iron, forms the important molecule of chlorophyll, the biomaterial in the green leaves of plants and in photosynthetic bacteria. Whereas chlorophyll has its most interesting and important biological properties when acting on its own account, the heme group has its most important properties when added to a chain of amino acids, a so-called polypeptide.

[2] *The Principles of Insect Physiology,* Chapman and Hall, London, 1972.
[3] *Nature,* 24 February, 1945.

The behaviour of the heme group depends markedly on the details of the chain of amino acids to which it is attached. For the polypeptide known as cytochrome *c* oxygen stays attached to the heme group, which in this case performs the function of giving electrons to other molecules. For the polypeptide known as myoglobin, whose structure was first worked out in detail by J.C. Kendrew, the heme group gives its oxygen to other molecules. Myoglobin is a hander-on of oxygen, absorbing it in one place (the lungs) and delivering it in another place (the muscles).

Hemoglobin, whose detailed structure was first worked out by M.F. Perutz, is also a hander-on of oxygen. Hemoglobin is a multiple polypeptide, with a heme group attached to each of several polypeptides. In adult human blood, for example, amino acid chains known as α and β, two αs and two βs, form into a multiple polypeptide with the four heme groups bonded into a tetrahedral structure. In the blood of the human fetus, on the other hand, the two βs are replaced by two γs, the γ chain being a third form of polypeptide. A fourth form, the δ chain, is also present in human genetic material (genome).

The hemoglobin of legumes is also built from four polypeptide chains, which are all distinct, however, and have been named, *a, b, c* and *d*. The *a* and *c* chains are comparable in their lengths with the globin chains of the higher animals, but *b* and *d* are significantly shorter. A detailed consideration of the legume *a* chain has suggested affinities with the human γ chain.[4]

The crucial property of both myoglobin and hemoglobin, of taking up oxygen and of then handing it on, a reversible reaction with respect to oxygen, is almost unique. It is not shared by organic molecules outside biology. Thus biology appears to have 'found' all the possible arrangements of atoms, highly complex arrangements totalling of the order of 10,000 atoms in the case of hemoglobin, that are capable of transferring oxygen molecules reversibly and efficiently. It will be hard enough, as we shall see later, to understand how animals with a decisive use for hemoglobin have been able to 'find' such an exceedingly complex structure, but how could peas and beans have found it?

In the nineteenth century and in the first half of the present century, before the amazing complexity of such molecules was revealed by modern microbiology, it was possible for the Darwinian theory to take refuge from this kind of question by the following argument. There have been very many generations of peas and beans, and in each generation there have been very many individual plants. Hence an exceedingly

[4] N. Ellfolk, *Acta Chem. Scand.*, vol. 13, 1959, 596; *Endeavor*, vol. 31, 1972, 139.

improbable event like the emergence of hemoglobin could happen to a few individuals and could then spread itself by natural selection, provided the happening conveyed some suitable advantage to the descendents of the few individuals that acquired it initially.

The style of this argument shows why for so long Darwinism was such a difficult theory to grapple with. The smallness of the probability of an important new biomolecule like hemoglobin emerging in any one individual was admitted, but was then said to be overwhelmed by the enormous assembly of individuals to whom the improbable event might happen. In the absence of reasonably firm estimates from the various numbers one was obliged to agree that it might have been so, and because of this unavoidable admission the critic was reduced to impotence. If indeed there were any critics at all. Most scientists took the correctness of Darwinism to be axiomatic, and they simply argued that any hypothesis needed to make Darwinism work had to be true.

Nowadays this situation is changed. The basic structures of the important biomolecules are mostly known and calculations of their being formed through shufflings of the constituent atoms can be made. It turns out that a successful shuffling is a vastly more unlikely business than looking for the tiniest needle in the largest haystack. So far from the number of legume plants which have existed on the Earth in the past being adequate to find hemoglobin, it looks as if hemoglobin could scarcely be found even if the whole universe consisted of peas and beans.

Twenty kinds of amino acids appear in the polypeptides which form hemoglobin and in most other bioproteins. There is a total of 141 amino acids in the human α chain and 146 in the β chain. The number of ways in which chains that are 141 amino acids long can be made up from twenty different kinds is $(20)^{141}$, vastly more than the number of atoms present in all the stars and galaxies that are visible in the largest astronomical telescopes. The funtioning of hemoglobin, however, does not demand a uniquely specified chain. A considerable fraction of the $(20)^{141}$ possibilities will do more or less equally as well as each other. This permits what are called amino acid 'substitutions' to occur without deleterious effects to creatures that depend on hemoglobin. Thus some, but by no means all, of the amino acids along the α and β chains can change their identities without serious effects. As a very general estimate, some twenty to thirty such substitutions are possible in the α chain, for example, without more than moderate changes taking place in the properties of hemoglobin, and perhaps another fifty substitutions are possible without destroying the basic function of the hemoglobin molecule.

As each generation is followed by the next generation, a copying of the genetic information present in the earlier generation takes place. The copying process is nearly one hundred per cent accurate, but not quite. The occasional inaccuracies, whose frequency we shall estimate in a moment, lead to the identity of particular amino acids being changed along the α and β chains. There are similar occasional changes in every bioprotein, for example in cytochrome *c*.

These random copying errors mostly lead to bad effects, and the individual recipients of seriously deleterious errors are so handicapped that they fail to reproduce themselves, so that the error does not carry through into later generations. This is prohibitive natural selection, the preventing of serious mistakes in our genetic heritage that was first noticed in 1835 and 1837 by Edward Blyth.

The opposite can also happen, a random copying error can lead to an appreciable improvement. The now-lucky recipient of the change is then better able to survive and to pass on the improvement to the next generation. So the improvement tends to spread through the positive natural selection that is the basis of Darwinism. However, finding improvements to such a highly sophisticated molecule as hemoglobin is like monkeys on a typewriter finding improvements to Shakespeare. It can happen only very rarely, and the question is whether it can happen often enough to give Darwinism any hope of being able to explain the evolution of plants and animals. Long before the publication of Darwin's book *The Origin of Species* in 1859, Edward Blyth had considered the problem in some detail. He decided there were insuperable objections against the origin of species being determined by what he called 'erratic' changes. Since Blyth's work in 1835 and 1837 has been largely forgotten, and since we believe his arguments to have been more correct than those of Darwin, we discuss the matter at some length in a historical note at the end.

Far more frequently than advantageous random changes, but less frequently than deleterious ones, there will be changes that are irrelevant, so-called neutral mutations. Some geneticists object that there can be no such thing as a neutral mutation, that a mutation must always convey some advantage or disadvantage, however slight. This, however, is not the issue. The issue is whether the advantage or disadvantage conveyed by a mutation is sufficiently large to be of relevance in comparison with other random effects that are always taking place within every breeding group of plant or animal. A parent may possess a variation of a particular gene on a particular chromosome, but unless the advantage or disadvantage conferred by the mutation in question happens to be large enough the random effect which inevitably occurs in the mixing of male and female chromosomes during the production of sex cells (gametes) overwhelms the effect of the mutational variation. A mathematical discussion shows that this stochastic effect (i.e., an effect governed by the laws of probability) is more important for small breeding groups than for large ones. For any specified breeding group, the stochastic effects define a level of advantage or disadvantage below which mutations can be considered to be effectively neutral. As we have remarked already, this class of neutral variations is far more numerous than the class of significantly advantageous mutations. For the purpose of our present discussion, some twenty to thirty variations of the amino acids on the α and β chains of hemoglobin can perhaps be considered neutral.

Because of the neutrality of some amino acid variations, different animals have moderately different α and β chains. Very strikingly, the differences are far smaller when two species are close to each other in the evolutionary sense than when they are widely separated. Man's close relative, the gorilla, has only one difference in its α chain (from the human α chain) but the horse has eighteen differences and the rabbit has twenty-five differences.[5]

The concept implied by 'evolutionary sense' in the previous paragraph is that all the mammals were derived from a common ancestor. The ancestral stock forked into two or more branches, each of which then forked again, and so on. The nearer two species are to each other in this 'ancestral tree', the closer their relationship in an 'evolutionary sense'.

The gorilla and man are thought to have diverged in the tree only about seven million years ago, while man and the horse diverged perhaps seventy million years ago. It is not the time intervals along the branches of the ancestral tree that are relevant to the accumulation of neutral mutations, however, but the number of generations which elapsed along the branches. A creature with a short generation time like the rabbit or mouse accumulates neutral variations more rapidly than a long-lived creature like man. The fact that the rabbit, in spite of its short generation time, has not many more variations from man in its α chain than the horse suggests that no more than about thirty amino acid changes along this particular polypeptide chain can be considered neutral. Other changes besides the ones we find today must surely have occurred but were deleterious, and so were removed by prohibitive natural selection.

Creatures other than the mammals have wider ranges in their α-globin chains. Thus the carp has sixty-eight variations in its α chain from

[5] M. Kimura and T. Ohta, *Berkeley Symp. Math. Stat. Prob.*, 1972, vol. 5, 43.

man, indicating that changes probably deleterious to mammals are not deleterious to fish.

A similar pattern is shown by other biomolecules, for example by cytochrome *c*. Man has only one variation in cytochrome *c* from the monkey, but seventeen from the horse. On the other hand, the horse has only one variation from the donkey. Man also has seventeen variations from the duck, but the duck has only three from the pigeon, three from the chicken and three from the penguin.[6] Once again, creatures close to each other in the ancestral tree show few variations, while creatures farther apart show many variations. As with hemoglobin, the number of variations does not increase beyond a certain point among the mammals. Man has even more variations from the horse than from the rabbit (seventeen against twelve) and as many from the horse as from the duck. This indicates that not more than about fifteen variations are neutral in cytochrome *c*, at any rate for mammals.

Although it is high time we returned to the problem of peas and beans, it is worth continuing a little further in order to pick up an important point, which follows from putting together two remarks made above. Man diverged from the gorilla about seven million years ago, and in that time only a single neutral mutation has appeared to make a difference between the α-globin chains of gorilla and man. If we take the average generation time of our human ancestors to have been twenty-five years, and the average generation time of gorilla ancestors to have been ten years, the sum of the number of human generations and gorilla generations that have occurred since the divergence seven million years ago is about one million.

If the α-globin chain had only a single permitted neutral amino acid variation, namely the one which actually occurred at the 23rd position on the chain (glutamic acid in man, aspartic acid in the gorilla), the chance of a DNA copying error causing a change of any particular amino acid would be about one in a million per generation. In the case of the α chain there are probably some twenty to thirty neutral changes possible, however, and any one of them might have happened in the total of one million generations since man and gorilla diverged on the ancestral tree. The chance of a copying error is therefore some twenty to thirty times smaller than we have just calculated, about one in twenty to thirty million per generation.

This is a very slow rate. Yet it is only this very slow mutation rate that Darwinism has to work with. It is not only inadequate to explain the

[6] W.M. Fitch and E. Margoliash, *Science*, vol. 155, 1967, 279.

evolutionary changes that have occurred, sometimes over quite short intervals for both plants and animals, it is woefully inadequate.

Thus suppose one considers the evolution of only a single gene, and suppose as a minimum requirement for the gene to be suitably functional that only ten amino acids in the associated polypeptide chain have to be of exactly the correct kinds in exactly the correct positions. Forgetting the other non-critical positions, think of the ten critical ones as being numbered sequentially. By hypothesis, the correct amino acids are not at these ten positions to begin with. It will therefore take twenty to thirty million generations to change the amino acid at the first position, and then with only a small chance of the change being to the required amino acid. Suppose, generously, that the first position happens to change correctly in one hundred million generations. The other nine positions are likely to have also experienced changes during these one hundred million generations, but by no means changes to amino acids of the required identities. Indeed the chance that when the first one falls right all the other nine will also be exactly right is of the order of one part in $(20)^9$, because there are twenty kinds of amino acid to choose from at each of the other nine positions. In another hundred million generations we may think to get the amino acid at the second position right, and then in a further hundred million generations to get the third position right, and so on. We are doomed to disappointment, however, because during the hundred million generations in which we are getting the second position right the first amino acid jumps away and becomes wrong again. The situation is like a plumber's nightmare—no sooner is a leak repaired in one place than another starts up somewhere else. It is not hard to see that of the order of $(20)^{10}$ million generations are needed before the right ten amino acids fall into the right ten positions, and at last just one needed polypeptide becomes available.

This calculation shows that the legumes did not happen on their hemoglobin by chance. Either they had the hemoglobin from the 'beginning' or they acquired it from somewhere else. It is relevant to a discussion of these possibilities that investigators have not obtained hemoglobin from normal legume cells, but from the root nodules of plants that were invaded by the nitrogen-fixing bacterium *Rhizobium*. The association with the bacterium permits the DNA of the legume to 'express' globin genes that normally go 'unexpressed'. It will be recalled that a gene is a blueprint for producing a polypeptide, an explicit chain of amino acids. An expressed gene is one that actually gives rise to the corresponding polypeptide at some stage in the normal operation of a cell, whereas an unexpressed gene lies dormant in normal operation with no corresponding polypeptide appearing.

Most DNA goes unexpressed. It has been estimated that between ninety and ninety-five per cent of human DNA is unexpressed. The 'silent majority' of our genetic heritage almost surely plays a role in regulating the activities of the cell, but it is doubtful that regulation needs more than a small fraction of the large amount of silent DNA. At first sight therefore our cells would seem to be carrying a wastefully great quantity of unused genetic material. It is a tenet of Darwinian theory that unused material, material untested by natural selection, must be nonsense material. Such material would be incapable, if activated, of immediately yielding new proteins of relevance to the cell. The case of hemoglobin in legumes suggests this view to be wrong. A normally unexpressed set of genes when triggered by *Rhizobium* turn out to yield, not just a tolerably sensible product, but an exceedingly sophisticated product.

The mystery of the presence of hemoglobin in legumes would be alleviated and the Darwinian theory saved on this particular score if it could be shown that the relevant genes had been acquired from some other source. Genetic cross-connections exist between different strains of bacteria. The main genetic material of a bacterium is in the form of a closed loop. As well as its main loop (chromosome) a bacterium often contains much smaller subsidiary loops (plasmids). Plasmids can pass from one bacterium to another, even between different species of bacteria. Since, moreover, a plasmid can both form out of the main chromosome and add itself into the chromosome, bacteria possess a powerful means for exchanging genetic information.

Since *Rhizobium* has been found to contain genes for hemoglobin, it might be argued that genetic information for producing the relevant polypeptides was originally passed as a plasmid from the bacteria to the legumes. The hemoglobin of *Rhizobium,* however, is not the same as that of the legumes.[7] A second objection to such an argument is that the genes of bacteria are now thought to be fundamentally different from the so-called split genes of eukaryotic cells (of which the legumes are composed). A third point is that while the symbiotic relation of the bacterium to the plant might at first sight seem highly relevant, other cases are known in which bacteria not in close association with eukaryotic cells still generate similar complex bio-substances.

The pterins are pigments that were first isolated from the wings of butterflies and from the covering material of insects. In 1949, M. O'L.

[7] C.A. Appleby, *Biochim. biophys. Acta*, vol. 172, 1969, 88.

Crowe and A. Walker reported the isolation of a similar pigment from the tubercle bacillus, similarity being determined by spectroscopic analysis, fluorescence characteristics and chromatographic behaviour.[8] These authors also obtained pterin-like pigments from the diptheria bacillus, from mammalian tissue and from human and animal urine. This seems a clear case of the genes for the production of the pigment being spread widely across the whole face of biology. A similar possibility suggests itself for the globin genes in legumes.[9]

In logical support of the possibility that the globin genes in legumes are specific to the legumes themselves, nothing very substantial would be achieved even if it could be proved that they were derived by plasmid transfer from the bacteria, for we would still need to account for their presence in *Rhizobium*. It is all very well to explain the appearance of unexpected genes in unexpected places by a transfer process from some other place, but at some stage the origin of the original gene(s) must receive explanation. In view of the probabilities calculated above, one can see the difficulty plainly, a difficulty enormously reinforced in the following chapter, and a difficulty well brought out by our third signpost.

After the second world war, people in general became conscious of the need to sterilize canned foods—having suffered too much from unsterilized stuff during the war itself. The question arose of whether meat and fish already packed in cans could be sterilized by subjecting them to large doses of X-rays. To the surprise of bacteriologists it was found that the method was unworkable, because hitherto unsuspected kinds of bacteria could withstand enormous X-ray doses, many millions of times greater than they could ever have received in the natural environment. To have employed such enormous doses, besides being dangerous to humans, would have been uneconomic for the food-canning industry.

The bacteria in question were of an approximately spherical form, micrococci, and they became known as the radioresistant micrococci. We shall discuss them in more detail in chapter 3. Here all we need notice is that their ability to withstand X-ray doses was not a passive quality. It was not due to their remaining undamaged. They were heavily damaged. Their ability lay in a capacity to repair the damage with astonishing

---

[8] *Science*, vol. 110, 1949, 166.
[9] Although the heme group attachments to the resulting polypeptides may possibly be derived from the bacterium as discussed by C.A. Godfrey, D.R. Coventry and M.J. Dilworth, in *Nitrogen Fixation by Free-living Micro-organisms*, ed. W.D.P. Stewart, Cambridge University Press, 1975.

efficiency. The question was, and still is, how the highly specific enzymes needed for the repair process came to be present in the bacteria, since in the terrestrial environment there are no X-rays that could exert selective pressure for their development.

Attempts were of course made to find some other selective process which happened by chance to produce the same kind of pressure that X-rays would have done. If one believes in Darwinism there *has* to be some other such process, and because scientists believe in Darwinism there is a strong social tendency in this kind of situation for everybody to become satisfied with a weak explanation. Even so, no explanation has been found to be acceptable in this case. The issue remains an apparent mystery, as we shall find in chapter 4.

This example of an unearthly property is akin to insects responding to ultraviolet light at 2537 Å. Both insects and the radioresistant micrococci can deal with situations that do not exist at the surface of the Earth, and which have never existed on Earth unless perhaps in its very early history. Some would argue, implausibly we think, that the bacteria are 'remembering' a property which developed by natural selection about four thousand million years ago. The same argument cannot be used for insects, however, because the insects did not exist until about four hundred million years ago, and by then there were certainly no X-rays or 2537 Å ultraviolet at the Earth's surface.

The indications for insects, peas and beans, and micrococci are the same, that they possess genes with capacities which cannot be related to their respective environments, as if the genes were derived from some kind of pre-existence. As we proceed from chapter to chapter we shall find the evidence for this point of view will become stronger and stronger. The cases we have considered here are but a small beginning to a very big story.

2
Enzymes and other biochemicals

It was argued in the preceding chapter that the usual theory of mutation and natural selection cannot produce complex biomolecules from a random association of atoms—i.e. from a disordered chemical situation. We must expect therefore that many important biomolecules will be found which run across the whole face of biology without there being any evidence in the usual evolutionary record of their ever having originated.

This expectation is amply borne out by observation. In particular, the enzymes are a large class of molecule that for the most part runs across the whole of biology, without there being any hint of their mode of origin. There are about two thousand of them. Enzymes are polypeptides (proteins) that specialize in speeding up biological reactions, which they do with far greater efficiency than man-made catalysts. They act both to build up and to break down a wide range of biosubstances.

The surface shapes of enzymes are critical to their function. Figure 2.1 illustrates one step in a complicated series of reactions that serve biochemically to release energy from glucose. The enzyme shown here schematically is hexokinase. The effect is to add a phosphate group P to the glucose, taking it from adenosine triphosphate, ATP. Provided another phosphate group is available in the environment, the resulting adenosine diphosphate, ADP, returns to ATP: ADP + P → ATP. The reaction of Figure 2.1 is then repeated for a second glucose molecule. So the same enzyme molecule can assist the energy release from many glucose molecules. Although the figure indicates very well the importance of the shape of the enzyme to the specific reaction which it promotes, because the reaction takes place in three dimensions the situation is actually far more specific than a diagram in two dimensions can possibly illustrate.

Surface shape is therefore all-important to the function of an enzyme. Surface shape is determined by the particular sequence of amino acids in the polypeptide structure. One can think of getting the surface shape right in two stages of approximation. There are some ten to twenty distinct amino acids which determine the basic backbone of the enzyme and these simply must be in the correct position in the polypeptide structure. The rest of the amino acids, usually numbering a hundred or more, then control the finer details of the surface shape. There are also the active sites that eventually promote the biochemical reactions in question, and these too must be correct in their atomic forms and locations.

Consider now the chance that in a random ordering of the twenty different amino acids which make up the polypeptides it just happens that the different kinds fall into the order appropriate to a particular enzyme. The chance of obtaining a suitable backbone can hardly be greater than one part in 10^{15}, and the chance of obtaining the appropriate active site can hardly be greater than one part in 10^5. Because the fine details of the surface shape can be varied we shall take the conservative line of not 'piling on the agony' by including any further small probability for the rest of the enzyme. The two small probabilities we are including are quite enough. They have to be multiplied, when they yield a chance of one part in 10^{20} of obtaining the required enzyme in a functioning form.

By itself, this small probability could be faced, because one must contemplate not just a single shot at obtaining the enzyme, but a very large number of trials such as are supposed to have occurred in an organic soup early in the history of the Earth. The trouble is that there are about two thousand enzymes, and the chance of obtaining them all in a random trial is only one part in $(10^{20})^{2000} = 10^{40,000}$, an outrageously small probability that could not be faced even if the whole universe consisted of organic soup.

If one is not prejudiced either by social beliefs or by a scientific training into the conviction that life originated on the Earth, this simple calculation wipes the idea entirely out of court. But if one is so prejudiced it is possible, in the fashion of a grand master with a lost game of chess, to wriggle ingeniously for a while. He would make a series of postulates (for which there is no evidence) in the following way.

Suppose at each place where a wanted enzyme happened to arise by

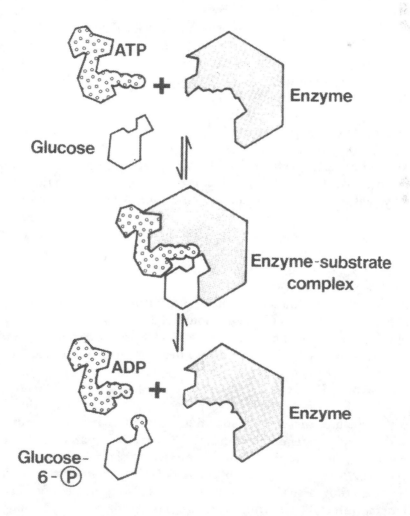

Fig. 2.1 The formation of an enzyme substrate complex, followed by catalysis (Adapted from J.D. Watson, *Molecular Biology of the Gene,* W.A. Benjamin Inc., 1977)

chance that the one enzyme alone formed part of a self-replicating chemical system, so that two thousand different self-replicating systems arose in different localities. Then each enzyme might be produced in large quantity in its own locality. To begin with, the different localities would be separate, but if the quantities produced by the self-replicating systems became large enough the two thousand initially different places might begin to overlap each other, finally bringing all the enzymes together, as required.

What has happened in the argument is that the minuscule probability of one part in $10^{40,000}$ has become buried in two thousand separate bits, each bit being the chance of a separate wanted enzyme just happening to fit into a separate self-replicating system. The weakness of the position is that such a hypothesis has only to fail in a few cases and the argument collapses. Thus if in a particular instance five enzymes were required to operate in concert with each other in order to promote a self-replicating system, the chance of such a system arising would be of the order of one part in $(10^{20})^5 = 10^{100}$, which is itself so small as to wipe out the idea. No slack in the argument is permitted at all. Even the need for only two enzymes to operate in association is sufficient to make the situation quite implausible.

Another way one might seek to argue would be to assert that not all the enzymes are independent of each other, as was assumed in the above calculation. This is true in some cases. The enzymes trypsin and chymotrypsin are evidently related to each other and were very likely derived from a common source. But cutting down the number of independent enzymes from two thousand does no good at all, unless the number were cut to a mere one or two, and this is not a viable suggestion. Besides which, just the same difficulties arise from a consideration of other cases. For the biomolecules known as histones the argument is unequivocal. It is worth a diversion to emphasize this point.

The histones are also polypeptides. There are only a few of them, with structures well enough known for us not to have trouble ourselves with the independence question of the previous paragraph. There are five histones (some investigators have described a sixth) referred to as H1, H2A, H2B, H3 and H4. Except at times of cell division DNA is not easily observed, but at cell division the DNA stands out much more sharply, looking rather like a string of beads. The beads consist of about two hundred DNA base pairs wound in a 'superhelix' onto a core of H2A, H2B, H3 and H4. In the lengths between the beads, the DNA is wound or looped over H1. The composite material of DNA and the histones is referred to as chromatin. Likely enough, this structure is temporarily adopted to give sufficient support to the DNA to enable it to withstand the mechanical rigours of cell division.

The shortest histone polypeptide is H4 with rather more than a hundred amino acids, and H1 with rather more than two hundred is the longest. Their function must be considerably more subtle than merely structural, because cells are evidently very 'fussy' in making sure they have the histones dead right for their needs. Thus the genes which code for the histones exist on the DNA in adjacent (tandem) clusters, with several hundred gene copies in each cluster. Moreover, the histone amino acid sequences are remarkably invariant across the whole of biology. Thus H4 has only two amino acid variations between cattle and the garden pea. To obtain what is an essentially unique sequence by random choices from equal quantities of twenty kinds of amino acid would require of the order of $(20)^{100}$ trials (there being about a hundred amino acids in the sequence). Once again we have a number larger than the total of all the atoms in all the stars and galaxies visible in the largest astronomical telescopes. This is for H4 alone, so there is no question here of whether or not a group of biomolecules are independent of each other.

Although our losing grand master now has to face open files and diagonals around his king, he can wriggle a little further. The histones have rather high proportions of the amino acids lysine and arginine, so one could shade the probabilities a little (of their forming by random associations of amino acids) by supposing them to arise in a 'soup' that was especially rich in lysine and arginine. Very well then, let us consider the t-RNA molecules.

The t-RNA molecules supply the last link in a complex process whereby DNA codes for a polypeptide. It is the function of each t-RNA molecule to interpret one or more particular base triplets, derived originally from the DNA, in such a way as to deliver a particular amino acid. There are many kinds of t-RNA molecule, and it is the sum total of their properties as a set that determines the well-known genetic code.

With minor exceptions, the genetic code is 'universal', which means that it goes across the whole face of terrestrial biology. This means that all plants and animals, and all bacteria too, have t-RNA molecules that operate in exactly similar ways. It is not hard to appreciate that the genes on the DNA which code for the structures of the t-RNA molecules must have been held fixed by fierce prohibitive natural selection. Imagine that due to a copying error occurring to a t-RNA gene a rogue t-RNA molecule emerged with the property of delivering a wrong amino acid in correspondence with a particular base triplet. The rogue molecule would then operate to build the wrong amino acid into every polypeptide produced by the cell. For some polypeptides the mistake might not matter too much, but for others the mistake would fall at crucial places in the chains, thus rendering the cell inoperable.

The first detailed structure of a t-RNA molecule was published in 1965.[1] This was the particular t-RNA in yeast cells that delivers the amino acid alanine. The structure is shown in Figure 2.2, in which A, U, G, C are the usual bases, adenine, uracil, guanine and cytosine. Since A, U, G, C are themselves collections of atoms, as are the unusual bases shown in the figure, and since the actual three-dimensional structure of the molecule is considerably more complex than its two-dimensional representation, it is clear that the chance of arriving at such a structure by random processes must be very small, even if one were given the various bases to start with. Assigning an explicit probability for arriving at a highly complex structure like Figure 2.2 is not so clear-cut as it was for the simpler chains of amino acids considered above, but our grand master can gain little relief from this technical difficulty. The probability will assuredly be small, and it will moreover be raised to something like the 60th power, for the reason that there are of the order of sixty different t-RNA molecules.

Indeed nothing remains except a tactic that ill-befits a grand master, but which was widely used by staunch club chess players in the youth of one of the present authors (F.H.), namely to blow thick pipe tobacco-smoke into our faces. The tactic is to argue that although the chance of arriving at the biochemical system of life as we know it is admitted to be utterly minuscule, there is in Nature such an enormous number of other chemical systems which could also support life that any old planet like the Earth would inevitably arrive sooner or later at one or another of them.

This argument is the veriest nonsense, and if it is to be imbibed at all it must be swallowed with a jorum of strong ale. The two commonest molecules in the universe are hydrogen, H_2, and carbon monoxide, CO. Joining these two gives formaldehyde, H_2CO, and joining six formaldehyde molecules into a suitable structure gives glucose. Thus the basic way in which terrestrial biology obtains energy, through the fermentation of glucose, is not a mere face in the crowd, a crowd of $10^{40,000}$ possible chemical systems. The use of glucose is a universal highway. Nor is there a profusion of chemical systems in which glucose can be used, otherwise chemists would long ago have found at least some of

[1] R.W. Holley, J. Apgar, G.A. Everett, J.T. Madison, M. Marguisee, S.H. Merrill, J.R. Penswick and Z. Zamir, *Science*, vol. 147, 1965, 1462.

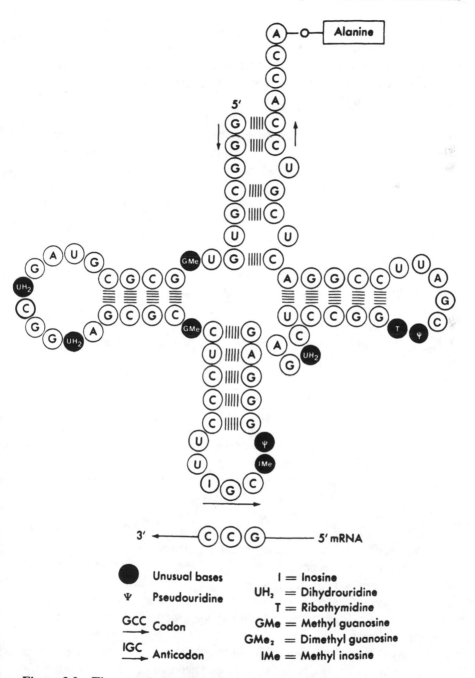

Figure 2.2 The complete nucleotide sequence of alanine t-RNA showing the unusual bases and codon/anticodon position (Adapted from J.D. Watson, *Molecular Biology of the Gene*, W.A. Benjamin Inc., 1977)

them. And if obtaining energy from glucose were such an easy matter it would long since have been found possible to circumvent the very complex sequence of reactions whereby the energy is actually yielded in terrestrial biology. So far from there being very many indistinguishable chemical possibilities, it seems that we have an exceedingly distinguishable system, the best.

Even with all this devastating evidence against the usual point of view, our discussion so far is still quite peripheral to really explaining the origin of life from a terrestrial organic soup of bases, amino acids, phosphates, ... Nothing has been said of the origin of DNA itself, nothing of DNA transcription to RNA, nothing of the origin of the program whereby cells organize themselves, nothing of mitosis or meiosis. These issues are too complex to set numbers to. Likely enough, however, the chances of such complexities arising from a soup initially without information, a soup that can only proceed by trial and error, are still more minute than the exceedingly small probabilities estimated above.

For life to have originated on the Earth it would be necessary that quite explicit instructions should have been provided for its assembly. A theory coming roughly within these terms was suggested some twenty years ago, half-jokingly by Tommy Gold. A space-ship landed on the Earth in its early days, scattering living cells, which have evolved and persisted ever since. So far as we are aware, this suggestion has not received much serious discussion in the halls of academe. Yet because it satisfies the need for explicit biochemical information to be supplied to the Earth, Gold's suggestion is less improbable by very many orders of magnitude than the usual organic soup theory. A more serious form of this idea was discussed in 1973 by Francis Crick and Leslie Orgel, once again with a far higher probability of being correct than the usual theory.[2]

One might conceive of the required information being contained within the laws of physics and chemistry themselves. In the very broad sense that life is a phenomenon belonging to the universe, so that its origin and evolution must be consistent with the universe and its laws, this concept is evidently correct. But this is not the point at issue. The question is whether the laws of physics and chemistry require that life should have originated here on the Earth, in what from a cosmic point of view is nothing but tiny patch of land on the edge of the small pond we call the terrestrial ocean. The issue is whether the laws require DNA to be

[2] F.H.C. Crick and L.E. Orgel, *Icarus*, vol. 19, 1973, 341.

formed, the t-RNA molecules, the histones and all the amino acid sequences of two thousand enzymes—formed explicitly—in small ponds with suitable temperatures and with suitable dispositions of the chemical elements, everywhere throughout the universe.

If we could prove an affirmative answer to this question we would have solved the problem of the origin of life on the Earth, but in doing so we would have come back to what in essence was the special creation theory of the early nineteenth century, the theory that Darwinism was supposed to have replaced. The only difference would be that the word 'laws' would replace the word 'God'. And it would be only a short step to say that God created the laws, and then we would be back to an identical position, except with more technical words in which to express it.

The obvious escape route is to look outside the Earth, although we should be warned that even this route may not be easy to follow. There is no way in which we can expect to avoid the need for information, no way in which we can simply get by with a bigger and better organic soup, as we ourselves hoped might be possible a year or two ago. The numbers we calculated above are essentially just as unfaceable for a universal soup as for a terrestrial one. The advantage of looking to the whole universe is rather that it offers a staggering range of possibilities which are not available here on the Earth. For one thing it offers the possibility of high intelligence within the universe that is not God. It offers many levels of intelligence rising upwards from ourselves. But before taking such 'far out' ideas seriously it is essential to become convinced that a route outside the Earth really is the correct one. A judgment on this matter must be based on fact. Fortunately, the facts of terrestrial biology are many and varied. It will be our aim in the next several chapters to interpret biology within the framework of a cosmic theory now to be stated.

Genes are to be regarded as cosmic. They arrive at the Earth as DNA or RNA, either as fully-fledged cells, or as viruses, viroids or simply as separated fragments of genetic material. The genes are ready to function when they arrive. Those for the enzymes, for example, will yield polypeptides with their amino acids in appropriate sequences to carry through all the catalytic reactions that we associate with the enzymes.

The problem for terrestrial biology is not therefore to originate the genes but to assemble them into whatever functioning biosystems the environment of the Earth will permit. This is not to say that copying-error mutations of the usual type cannot modulate the genes, but the effect will be a fine tuning capable only of distinguishing the varieties of a species. The copying-error mutations will not be capable of producing the broader divisions of the plants and animals.

The genes do not arrive at the Earth all in one moment during its early history, as in the suggestion of Tommy Gold. They arrive continuously but not necessarily steadily. There can be high peaks in the arrival rate, lasting for thousands and perhaps even for hundreds of thousands of years. Nor do the genes arrive in a science-fiction vehicle. They arrive without any vehicle at all. Large stores of genetic material became frozen and so preserved indefinitely in the outer regions of the solar system during the early days when our system was formed about 4.6 billion years ago. Comparatively small quantities of the genetic material are released continuously through the agency of comets in the manner discussed in two former books *Lifecloud* and *Diseases from Space*. But whereas we considered in those books that life might have arisen in the comets, we now regard the comets only as possible amplifiers of the numbers of pre-existing cells and as storage refrigerators for them.

This picture does not, as stated, necessarily imply a purpose, but it will simplify the discussion from here on if we state our clear-cut view that there is indeed a purpose. The purpose is to generate life, not just on the Earth, but everywhere that it will take root, and to do so in an extremely elegant way. Not by parading around the galaxy in a fleet of space ships, but through the genes travelling like seeds in the wind— more explicitly the genes ride around the galaxy on the pressure of light waves from the stars and they do so at speeds up to several hundred kilometres *per second*. (Genetic material is of exactly the right size to ride on the light waves of stars, and to do so with the greatest possible efficiency. No rocket engines are needed.) Unlike the crew of a space ship the genes are potentially immortal.

Our views on biology have been treated in a curious way. While our own inquiries, particularly in the medical field, have always been treated with the greatest courtesy, nobody has overtly sought us out like a Dutch uncle, patted us on the head, and told us to stop talking nonsense. Neither have we been attacked with the ferocity that we expected. It is rather that our writings have been greeted with a wall of silence. The reason very likely is that every expert biologist has seen from the beginning that sooner or later the word purpose would appear, and to involve purpose is in the eyes of biologists the ultimate scientific sin, worse even than to express doubt of the validity of Darwinism.

It would be acknowledged in scientific circles, except perhaps among extreme behaviourists, that humans often act with purpose. Apes clearly act on occasion with purpose, and so too in various degrees do others of the higher animals. As one descends the scale of electronic complexity among animals, however, purpose is usually considered to disappear, to be replaced by instinctive (through-wired) survival reactions. 'Purpose' is therefore a state of mind that goes with increasing brain complexity. Stated briefly, purpose is an attempt to control events in the world according to some consciously thought-out policy. The higher the intelligence the more effective purpose becomes.

The revulsion which biologists feel to the thought that purpose might have a place in the structure of biology is therefore revulsion to the concept that biology might have a connection to an intelligence higher than our own. This fits the way most people also think. While many are willing, and some are anxious, to postulate an ultimately surpassing intellect, God, few are happy with the thought of intelligences intervening at levels between ourselves and God. Yet surely there must be such intelligences. It would be ridiculous to suppose otherwise. Can there be intelligences not only with an interest in biology, but also with the ability to monkey with it, or even to control it? When one considers that we humans are already beginning to monkey with biology, in our so-called genetic engineering, the answer must surely again be yes.

Our postulate of a purpose in biology does not therefore seem to us to be the outrageous notion which it will probably appear to most biologists. In a similar way we think there could be purpose in astronomy, and perhaps even in cosmology. It all depends on where one sets the upper level of intelligence before reaching ultimately upwards to God. If indeed there is such an ultimate limit. The mathematician may prefer to think in terms of a sequence that never attains its limit, in which case *God* is an ideal element that exists in the abstract but which is never reached in practice. But this is to anticipate the final speculations of this book. For the moment our task is to consider how the ideas set out above deal with the practical details of terrestrial biology, and how they compare in their explanatory power with Darwinism.

3
The Panspermia theory

The theory outlined in the previous chapter has important correspondences with the so-called panspermia theory of Arrhenius, although in its implications for continuing biological evolution on the Earth it goes considerably outside the Arrhenius theory. We begin here with a description of that theory and with a short account of its history.[1]

Svante August Arrhenius was born at Vik, Sweden, on 19 February 1859, the son of a surveyor and land agent. He has been remembered more in scientific history for his work on electrolytes than for the development of the panspermia theory. It is a matter of everyday knowledge that if the liquid from a well-charged car battery is emptied and the battery is filled only with distilled water, little in the way of an electric current can be obtained from it. The reason is that pure water is a poor conductor of electricity. If, however, some acid—usually sulphuric acid—is added to the distilled water, the battery immediately delivers its normal current. The acid which thus plays a crucial role in transmitting electricity through water is known as an electrolyte. Common salt, which is made up of molecules each consisting of an atom of sodium and an atom of chlorine, is also an electrolyte.

As a student at Uppsala University, Arrhenius suggested in his doctorate thesis that electrolytes dissociate on solution in water. This means that the molecules come apart, as for instance the sodium atom separates from the chlorine atom in the case of molecules of common

[1] Adapted from our book *Space Travellers: The Bringers of Life*, University of Cardiff Press, 1981.

salt. The separation is of a peculiar kind, however, with the chlorine taking an electron away from the sodium. Arrhenius did not know about electrons at that time, 1883, but he suggested that when the molecules of an electrolyte dissociate into two pieces, the pieces are equally and oppositely charged, one positive, the other negative, and that it is the charging of the pieces which permits electricity to flow through the solution.

Correct as his theory was, the young student soon got into controversy. Disbelieved by most of the senior chemists in Sweden, his thesis received the lowest passing grade that the University of Uppsala could bestow. But Arrhenius seems to have been a stubborn fellow, not easily put down by this rebuff. He circulated copies of his work to leading scientists throughout Europe, and as the years passed the theory gradually became more and more accepted. Eventually it was judged respectable enough for Arrhenius to be elected a member of the Swedish Academy of Sciences.

Two years later, in 1903, he received the Nobel Prize for chemistry.

Having now come successfully through one scientific dogfight, Arrhenius almost immediately started another. This second controversy, beginning with the publication in 1907 of a remarkable book, *Worlds in the Making,* he was not destined in his lifetime to win.

Ironically, the ideas developed in *Worlds in the Making* come from the strict acceptance of one of the most important doctrines of nineteenth-century biology. It was Arrhenius who accepted the doctrine, and it has been his Earth-centred opponents who have persistently broken it. Louis Pasteur enunciated the doctrine when he told the French Academy of Sciences that the concept of spontaneous generation would not survive the 'mortal blow' which he had delivered to it by certain carefully designed experiments.

The concept of spontaneous generation, according to which life originates in mixtures of simple materials—earth, air, water—had persisted from Aristotle (384-322 BC) up to the middle of the nineteenth century. Nobody of course had seriously thought of big animals originating in this way. It was accepted that a calf can be obtained only from a cow, although in Shakespeare's *Antony and Cleopatra* Lepidus tells Mark Antony:

> 'Your serpent of Egypt is born in the mud, by the action of the Sun, and so is your crocodile.'

Small creatures—fire-flies, worms and maggots—were widely believed, however, to originate spontaneously out of simple inorganic materials.

Many had been the experiments that claimed to demonstrate the spontaneous emergence of life, but always when such experiments were repeated with better precautions the claimed results were shown to be incorrect. Already in 1668 the Italian physician Francesco Redi had shown that maggots would not form in meat if the precaution of keeping flies away from it was taken. More and more it became clear that in all cases where the spontaneous origin of life was supposed to happen, some form of living precursor, usually a tiny egg, was always present. Even the smallest creatures were like the calf and cow. Each generation of every animal was preceded by a generation of the same animal. This was the doctrine which Louis Pasteur enunciated more than a century ago to the French Academy of Sciences.

Yet by a remarkable piece of mental gymnastics biologists were still happy to believe that life started on the Earth through spontaneous processes. Each generation was considered to be preceded by a similar generation, but only so far back in time. Somewhere along the chain was a beginning, and the beginning was a spontaneous origin. Thus spontaneous origin was both true and it was false.

The contradictory nature of this reasoning was apparently softened by the Darwin-Wallace theory of natural selection. According to this theory, complex animals evolve slowly over the generations from simpler ones. In the beginning, the first animal could therefore be exceedingly simple, it was argued, and the spontaneous generation of an initially very simple creature did not seem so much of a problem as it would for the more complex creatures that we have around us today.

This argument, however, was illusory. 'Simple' animals may appear simple in their outward forms but in their internal chemistry they are highly complex. Thus the enormously complex molecules that we discussed in the two preceding chapters are required already for the so-called 'simplest' creatures. The biggest problem of biology lies in their origin, not in the evolution of the seemingly sophisticated animals. The recourse to an organic soup to cross this biggest hurdle is evidently a blatant recourse to the spontaneous generation theory which Pasteur claimed to have destroyed. Nevertheless, most scientists, even to this day, have been satisfied to accept it.

Most, but not all. Even in the nineteenth century there were a few scientists who felt the situation to be contradictory. If spontaneous origin could not happen, as Louis Pasteur had claimed to the French Academy, then it could not happen. Every generation of every living creature had to be derived from a preceding generation, going backward in time to a stage before the Earth itself existed. Hence it followed that life must have come to the Earth from outside, from a previous existence somewhere else in the universe.

This argument is quite powerful enough for it to have received respectful consideration by biologists and astronomers. There were of course subsidiary questions to be answered. How could life move from one planet to another? How could it cross the immensity of space from one star system to another?

The first attempts to answer these questions were crude. H.E. Richter, a German physician, argued that not all the chunks of iron and stone which enter the Earth's atmosphere from time to time (meteorites) plunge to the ground. Some meteorites must encounter the atmosphere at glancing angles, penetrating only a little way into the air, and then going out again, away from the Earth. Such meteorites might pick up living cells from the atmosphere, Richter argued, carrying them out into space to whatever destination the meteorites might eventually reach.

On a more ambitious scale, William Thompson (later Lord Kelvin) remarked in his presidential address to the 1881 meeting of the British Association:

> When two great masses come into collision in space, it is certain that a large part of each is melted; but it seems also quite certain that in many cases a large quantity of debris must be shot forth in all directions, much of which may have experienced no greater violence than individual pieces of rock experience in a landslip or in blasting by gunpowder. Should the time when this earth comes into collision with another body, comparable in dimensions to itself, be when it is still clothed as at present with vegetation, many great and small fragments carrying seeds of living plants and animals would undoubtedly be scattered through space. Hence, and because we all confidently believe that there are at present, and have been from time immemorial, many worlds of life besides our own, we must regard it as probable in the highest degree that there are countless seed-bearing meteoric stones moving about through space. If at the present instance no life existed upon this earth, one such stone falling upon it might, by what we blindly call *natural* causes, lead to its becoming covered with vegetation.

It is the remarkable merit of the final chapter of *Worlds in the Making* that it lifted these rather undeveloped ideas of Richter and Kelvin to the level of a serious scientific theory without immediate technical weaknesses.

The key to the thinking of Arrhenius lay in the phenomenon of radiation pressure. When light or heat is absorbed or reflected by any body, a physical pressure is exerted on the body. For large objects, radiation pressure is usually very small, but for small particles the pressure of radiation may be greater than any other force acting on the particles.

Arrhenius would seem to have had a poor opinion of most astronomers:

> The astronomers followed faithfully in the footsteps of their inimitable master, Newton, and they brushed aside every phenomenon which would not fit into his system. An exception was made by the famous Euler, who, in 1746, expressed the opinion that the waves of light exerted a pressure upon the body upon which they fell. This opinion could not prevail against the criticisms with which . . . others assailed it. That Euler was right, however, was proved by Maxwell's great theoretical treatise on the nature of electricity (1832).

Arrhenius then goes on to calculate the effect of the radiation pressure of sunlight on small particles in the solar system, correctly finding that there is a range of sizes in which the force of radiation pressure exceeds the force of solar gravity. Although eukaryotic cells, such as those which make up the human body, are outside this range on the high side (above about 3 microns) and viroids are outside it on the low side (less than about one-tenth of a micron) most bacteria and many viruses fall inside it. So bacteria and viruses, if in interplanetary space unencumbered by any surrounding material, would ride easily out of the solar system on the pressure of light waves from the Sun.

Arrhenius worries next about whether living cells expelled by radiation pressure from one star system could retain their germinating power for long enough to reach another star system. He answers confidently in the affirmative:

> When the spores have passed the orbit of Neptune, their temperature will have sunk to -220° (Centigrade) and farther out it will sink still lower. In recent years experiments have been made in the Jenner Institute, in London, with spores of bacteria which were kept for twenty hours at a temperature of -252° in liquid hydrogen. Their germinating power was not destroyed thereby.
>
> Professor Macfadyen has, indeed, gone still further. He has demonstrated that micro-organisms may be kept in liquid air (at -200°) for six months without being deprived of their germinating power. According to what I was told on the occasion of my last visit to London, further experiments, continued for still longer periods, have only confirmed this observation.
>
> There is nothing improbable in the idea that the germinating power should be preserved at lower temperatures for longer periods than at our ordinary temperatures. The loss of germinating power is no doubt due to some chemical process, and all chemical processes

proceed at slower rates at lower temperatures than they do at higher. The vital functions are intensified in the ratio of 1:2.5 when the temperature is raised by 10°C (18°F). By the time the spores reached the orbit of Neptune and their temperature had been lowered to -220°, their vital energy would, according to this ratio, react with one thousand millions less intensity than at 10°. The germinating power of the spores would hence, at -220°, during the period of three million years, not be diminished to any greater degree than during one day at 10°. It is, therefore, not at all unreasonable to assert that the intense cold of space will act like a most effective preservative upon the seeds, and that they will in consequence be able to endure much longer journeys than we could assume if we judged from their behaviour at ordinary temperatures.

Arrhenius does not know of the existence of cosmic X-rays, but he is aware that ultraviolet light from the Sun would have a potentially lethal effect on living cells:

> On the road from the earth the germs would, for about a month be exposed to the powerful light of the Sun, and it has been demonstrated that the most highly refrangible rays of the Sun (i.e. ultraviolet) can kill bacteria and their spores in relatively short periods. As a rule, however, these experiments have been conducted so that . . . [the spores were in a condition to germinate]. That, however, does not at all conform to the conditions prevailing in planetary space. For Roux has shown that anthrax spores, which are readily killed by light when the air has access, remain alive when the air is excluded. . . . All the botanists that I have been able to consult are of the opinion that we can be no means assert with certainty that spores would be killed by the light rays in wandering through infinite space.

For the next fifty years most biologists disgreed with this optimistic opinion. Not only would ultraviolet light prove lethal to Arrhenius' 'spores', but the discovery of cosmic X-rays—especially the occasional intense bursts of X-rays from the Sun and other stars—appeared to be an overriding objection to the panspermia theory. As late as 1960, this view was confidently held by nearly all biologists and astronomers.

By the middle 1960s it was realized, however, that graphite, which absorbs ultraviolet light over the wavelength range of outstanding importance for biology (around 2600 Ångströms), is present in great quantity in interstellar space. Roughly a third of all interstellar carbon is in the form of graphite. A layer of graphite only a few tenths of a micrometre thick provides effective shielding against harmful ultraviolet light. It is, moreover, a remarkable provision of Nature that when biomaterial becomes degraded in the absence of free oxygen, graphite is the end product.

In past ages here on the Earth, large quantities of biomaterial became suddenly buried from time to time, and so were shielded from atmospheric oxygen. This material then degraded more and more as time went on. It is of course the stuff we call coal. Various forms of coal are classified according to the extent to which they have lost their volatile materials. The highest rank of coal corresponds to the most advanced stage of degradation, with the most free carbon and the least of volatile materials. Coals with less than 8 per cent of the latter are termed anthracite, of which there is a special type containing less than 2 per cent of volatiles, and this is actually called graphitoid coal.

Thus, the natural product of biodegradation (under non-oxidizing conditions) is the very material that provides effective shielding against ultraviolet light. Through the destruction of a little of itself, a large-scale biosystem can therefore become automatically self-protecting against ultraviolet light. This appears to be exactly what has happened throughout our galaxy. A discussion of this question is given in detail at the end in the Technical Note. It is sufficient here that the presence of graphite, or indeed of any other form of free carbon (soot for example), is sufficient to dispose of the ultraviolet-damage problem that faced Arrhenius. Provided living cells are a little dirty with carbon at their surfaces they could even approach stars like the Sun, as they would need to do in order to reach the Earth from outer space.

When in the early 1960s the U.S. National Aeronautics and Space Administration (NASA) first began to think of sending a mission to Mars, specifically designed to test for the presence of life there, the question of terrestrial contamination had to be considered. Could living cells from Earth survive such a journey? If so, the Viking landers would need to be carefully sterilized before launch. If not, then the considerable expense of sterilization could be avoided. The fact that the landers were eventually sterilized when the *Viking 1* and *Viking 2* missions were flown in 1975-1976 shows what conclusion was drawn from the laboratory experiments that were carried out in the 1960s.

An early experiment was to test for the effect on micro-organisms of the exceedingly low pressure of interplanetary space. This would cause all free water to become dried out of the organisms, making cavities develop inside bacteria. Possibly this would lead to death and, if so, sterilization would be automatic, at any rate on the outer surfaces of the vehicles.

The outcome of such an experiment was reported in the issue of *Science* for 22 December 1961. It would scarcely be possible to express the results more clearly and briefly than the authors, D.M. Portner, D.R. Spiner, R.K. Hoffman and C.R. Phillips, did in their abstract:

> Three species of resistant micro-organisms were exposed for 5 days to an ultra high vacuum approaching that of interplanetary space. Since no lethal effect was observed, there is no indication that the vacuum of outer space would prevent transport of viable micro-organisms on unsterilized space vehicles.

This experiment was a warning signal. If the micro-organisms in question (*Bacillus subtilis, Aspergillus fumigatus* and *Mycobacterium smegmatis*) had originated on the Earth, where very low pressures are never encountered, how did it come about that they were nevertheless able to survive in such conditions? Surely it is a tenet of Darwinism that life-forms should not be well-adapted to environments which they have never experienced? Of course, if the micro-organisms were originally derived from space, the difficulty would not arise.

Here was another of Nature's signposts. Although it was largely ignored, a few thoughtful commentators began to write more guardedly about previous condemnations of the Arrhenius theory, for example, Carl W. Bruch in an article entitled 'Microbes in the Upper Atmosphere and Beyond'.[2]

> In view of the dormancy of micro-organisms in high vacuum and of their relatively low cross-section for ionizing radiation, the hazards to exposure in space may have been exaggerated. The chief danger to a microbe should come from solar ultraviolet radiation and the proton wind (also from the Sun), but a thin layer of overlying material would shield a spore from both types of radiation. Micro-organisms ejected from planets at great distances from a star such as Uranus or Neptune in our solar system should encounter negligible radiation hazards. [Ejected much more plausibly from comets than from Uranus or Neptune—present authors' comment.] The tenability of the panspermia hypothesis for ejection from or arrival on to many planets cannot be rejected by radiation sensitivity arguments alone.

The remark in this quotation about 'ionizing radiation' refers mainly to the effect of X-rays, which generate electrons of considerable energy by the so-called photoelectric effect when absorbed inside cells. The resulting electrons may cause damage to DNA. If an energetic photoelectron happens to be unluckily directed it can cause a breakage in

---

[2] In P.H. Gregory and J.L. Monteith (eds), *Airborne Microbes, Symp. Soc. Gen. Microbiol.*, vol. 17, Cambridge University Press, 1967.

one of the two strands of the DNA double helix. In rarer cases a photoelectron may even break both strands of the double helix.

The lower-energy X-rays are much more damaging than those of higher energy, with a worst quantum energy of about 1 kilovolt, although at this energy there could still be partial shielding from the same kind of overlying material that gives protection against the ultraviolet. (For comparison, quanta of visible light typically have energies of 3 volts, while biologically damaging ultraviolet light is made up of quanta with energies of about 5 volts.) To avoid the complicating effects of shielding, we should perhaps consider somewhat harder quantum energies of two or three kilovolts. During large solar flares, the Sun emits X-rays at 100,000 or more times its quiescent rate. But since the Sun fluctuates markedly in its flare activity, sometimes going for months or even years without large ones breaking out, it would be possible to take a 'safe' line by arguing that viable micro-organisms approach the Earth from space (or go out to space from the Earth in the Arrhenius theory) at times when there are no flares on the Sun. This caution, however, is scarcely necessary.

One can calculate that at the maximum of a large flare a micro-organism in interplanetary space at the Earth's distance from the Sun would receive a radiation dose of from 1 to 10 rads per second. Although this dose rate is high, large flares decline appreciably from their soft X-ray maxima in only about 10 minutes, so that the accumulated dose for a large flare is limited to a few kilorads (1 krad = 1000 rad).

Figure 3.1 shows the X-ray dose-resistance of three strains of *Escherichia coli,* the well-known bacterium that assists us in the digestion of food. *Escherichia coli* could evidently withstand several large solar flares. Other bacteria and some eukaryotic cells can withstand vastly higher doses of X-rays than *E. coli,* and for these there would be no question of being adversely affected by solar X-rays during an approach to the Earth. Thus *micrococcus radiodurans* and *micrococcus radiophilus* can withstand doses upwards of 600 krad, while *amoeba* and *paramecium* can withstand doses in excess of 100 krad.

It has been estimated that in exposure to a dose of 600 krad *micrococcus radiodurans* must experience some 18,000 single-strand breaks of its DNA double helix.[3] Since such breaks would be lethal if left unrepaired, it is evident that *micrococcus radiodurans* must have an exceedingly efficient repair mechanism at its disposal. Repair is carried

---

[3] A.D. Burrell, P. Feldschreiber and C.J. Dean, *Biochemica et Biophysica Acta,* vol. 247, 1971, 38.

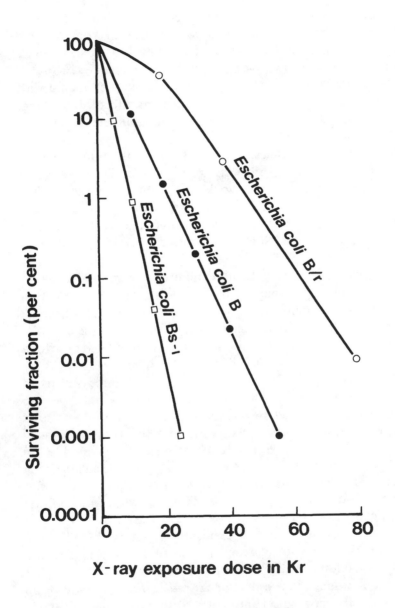

Fig. 3.1 Surviving percentage of three strains of *E. coli* as a function of X-ray exposure (From A. Nasim and A.P. James in *Microbial Life in Extreme Environments,* ed D.J. Kushner, Academic Press, 1978)

out by enzymes of which several appear to be involved. The strand break is recognized, bits are cut away on either side of it to form a small gap, and then the broken strand with the appropriate bases attached is repaired by making reference to the bases attached to the opposite undamaged strand. (It will be recalled that either strand of DNA with its attached bases can be used to generate the opposite strand with its attached bases, so building the correct double helix.)

As well as the 18,000 single-strand breaks it has also been estimated that there would be 1400 double-strand breaks. How these are repaired is less clear, but repaired they must be in order for the bacteria to maintain their viability.

There are no X-rays at the surface of the Earth, because solar X-rays (and the much weaker extra-solar system X-rays) are absorbed in the terrestrial atmosphere. Since, moreover, the Earth has had an atmosphere giving X-ray absorption for thousands of millions of years, for as long as there has been life on Earth, it is evidently a poser, not for the panspermia theory but for the conventional theory, to explain how micro-organisms acquired their enormous resistance to damage by X-rays. The very feeble background of ionizing particles coming from the low concentrations of radioactive materials present in rocks, water and soil is evidently minuscule compared to the intensely ionizing environment needed to provide sufficient selection pressure to explain the biological facts, and this is generally conceded to be so. What is usually said is that the situation is fortuitous. The battery of enzymes that repair X-ray damage in micro-organisms are supposed to have evolved in response to a different selection pressure and it is a matter of chance that they just happen to be so remarkably effective against X-rays.

If this were the only problem for the usual theory in its relation to radiobiology, the position would surely be weak enough, but there are others also, as we shall see in a moment. First, however, let us put the space theory through an important test. The abilities of micro-organisms to repair X-ray damage are essential for them to survive in space, where (as we saw earlier) the gas pressure is very low indeed compared with pressures on the Earth, with the consequence that free water becomes lost from the micro-organisms. Do they still survive damage from ionizing radiation under such conditions?

Work on this question was carried out with spores of *Bacillus megaterium* and results have been reviewed by E.L. Powers and A. Talentire.[4] In the absence of free oxygen, resistance to ionizing radiation

[4] In *Actions Chimiques et Biologique des Radiations*, vol 12, pp 3-67, Masson et Cie, Paris, 1968.

was improved slightly down to a pressure of about 1 per cent of a normal atmosphere, and thereafter resistance remained the same down to one millionth of a normal atmosphere. The indication therefore is that under conditions in space micro-organisms would at least retain their laboratory-measured abilities to survive damage from ionizing radiation. (Because of the exceedingly low concentration of gas in space, oxygen can be considered 'absent' there.) The test is passed satisfactorily as far as evidence is available.

We mentioned earlier that ultraviolet light with wavelengths around 2600 Ångströms is very damaging to living cells. Although a thin shield of free carbon would protect cells against ultraviolet light, it would be inconvenient, if not indeed impossible, to maintain such an outer layer of shielding material in all situations. When cells are active, and especially when reproductive, nutrient materials must enter them from outside. Thus access through cell walls and membranes is also necessary, and for this the outer layer of shielding material must be removed, so that cells then become susceptible to damage by ultraviolet light.

A possible resolution of this problem would be for micro-organisms to reproduce themselves only in highly special environments, where for some different reason a shield against ultraviolet light happened to be present. This very nearly does happen at the surface of the Earth, with the terrestrial atmosphere providing the shield. However, such a requirement would be highly restrictive. So in a powerfully driven cosmic system we would expect micro-organisms to possess highly effective repair mechanisms, not only against ionizing radiation (X-rays) but also against non-ionizing ultraviolet light, and this indeed is the case as we shall now see.

The absorption by DNA of ultraviolet light with wavelengths around 2600 Ångströms does not supply enough energy to break the DNA strands, but it causes changes in the chemical bonds within the double helix structure. The main changes occur between the so-called pyrimidine bases, thymine and cytosine. The effect is to establish linkages between the ring structures of these bases, when they happen to lie alongside one another in the DNA chain. There can be thymine-thymine (the most important), thymine-cytosine and cytosine-cytosine linkages as the case may be, which are referred to collectively as pyrimidine dimers. Less serious, but also relevant, there can be changes in the chemical bonds outside the ring structures, and linkages also can be established between DNA and nearby protein molecules.

It was discovered by A. Kellner in 1949 that pyrimidine dimers can be restored to their original base configuration through the exposure of micro-organisms to visible light. What happens is that a specific enzyme, different from those which repair DNA strand-breaks, binds to the site of a dimer. The enzyme-dimer complex then absorbs visible light, which supplies the energy needed to flick the chemical bonds back to their original arrangement. Thereafter, the enzyme detaches itself and proceeds to repair another dimer, and so on, the efficiency of the process being governed by the number of enzyme molecules available and by the time required for them to 'find' the dimers.

Unfortunately, laboratory experiments are carried out under conditions that are different from those that would apply in most natural situations. Micro-organisms are exposed in the laboratory to ultraviolet light and visible light sequentially, whereas in most natural situations both forms of light are present together. Thus in natural situations repair goes on simultaneously with dimer formation, whereas in the laboratory there is damage first and repair second. If the initial damage is extensive enough, the subsequent repair process may fail to restore an organism to viability. Figure 3.2 shows the fractions of the populations of various micro-organisms that were subsequently restored to viability after exposure to ultraviolet doses from a mercury arc lamp (2537 Ångströms), the doses being plotted along the horizontal axis of the figure. Different organisms evidently vary greatly in their resistance, just as they do in resistance to X-rays. The most ultraviolet-resistant organism known is the marine flagellate *Bodo marina,* which requires an ultraviolet dose of about 110,000 ergs mm^{-2} to destroy 90 per cent of its cells, a dose about twenty times larger than that needed to destroy the radioresistant micrococci.

The problem is to relate the laboratory results to the natural situation in which both ultraviolet and visible light are present together. This we have attempted to do, subject to some uncertainties in the data. Here we simply quote the results of our conclusions, namely that *E. coli* would be destroyed by the full space flux of ultraviolet and visible light (at the Earth's distance from the Sun) but that the radioresistant micrococci, many protozoa and *Bodo marina,* of course, could withstand the full solar flux.

This raises another difficulty for the conventional theory. Essentially no ultraviolet light with wavelengths less than 2900 Ångströms reaches the surface of the Earth, because of absorption in a layer of ozone at heights between about 25 and 50 kilometres in the atmosphere. Even at 3000 Ångströms only about 1 per cent of the solar near-ultraviolet reaches ground-level, and at wavelengths of 3000 Ångströms or more pyrimidine dimer-formation is inefficient—it is less than one-hundredth part as effective as at 2600 Ångströms. The combined effect of this weakened efficiency and of the still strong atmospheric absorption at

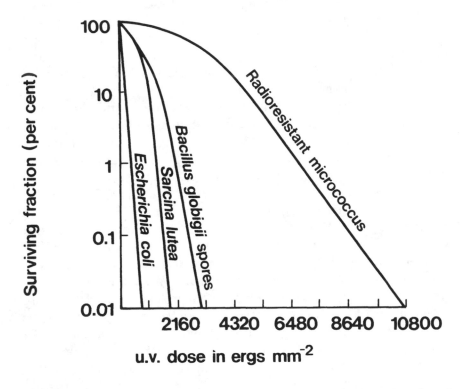

Figure 3.2 Surviving percentages for various micro-organisms as a function of ultraviolet exposure (A. Nasim and A.P. James, *op. cit.*)

3000 Ångströms is to reduce dimer formation at the Earth's surface to less than one-thousandth part of what it would be out in space. It follows therefore that in respect of conditions at the Earth's surface, the radioresistant micrococci and many protozoa are all grossly overprotected against ultraviolet light. The question for the usual theory is what selection pressure endowed them with this property.

The suggestion that much more of the solar ultraviolet penetrated the terrestrial atmosphere in the early history of the Earth, and that micro-organisms acquired their great ultraviolet resistance in those early days, does not sit well. It defies the evidence to suppose that such an early resistance could have persisted without selection pressure to maintain it, for a time span longer than 3000 million years, since experience shows that micro-organisms acquire radiation-sensitive mutations very readily.

This effect can be seen, for example, in the three strains of *E. coli* shown in Figure 3.1

Instead of presenting what are really quite insuperable difficulties, as in the usual theory, the facts concerning radiation resistance are easily explained by a cosmically oriented theory. Micro-organisms arrive at the Earth with great resistance against both ultraviolet light and ionizing radiation, because without such resistances they could not survive in the cosmic environment. Those organisms that manage to take root here at the Earth's surface are no longer exposed to the severe selection pressure of space. Mutants with weakened resistance are not therefore at a disadvantage and gradually replace the resistant strains as time goes on. What we are seeing in Figures 3.1 and 3.2 is this downhill process, with the resistant micrococci being more recent arrivals on Earth than *E. coli*, and with the three strains of *E. coli* in Figure 3.1 showing the degeneration process in operation for a particular species. Radiation resistance is therefore simply a measure of how long particular organisms have been here on the Earth. Hence, variable resistance is to be expected from species to species according to a terrestrially oriented roster of seniority. Variable resistance is not the impossible conundrum that it is in the usual theory.

A. Nasim and A.P. James write as follows:[5]

> Species vary enormously in their natural resistance to radiation. This variation far exceeds that expected from any differences in the mean levels of radiation to which organisms are exposed. There is, in fact, little or no correlation between natural levels of exposure and species resistance. The reasons for this variation are obscure, but it seems likely that extreme resistance is a consequence of some other feature of a species.

The 'other feature' is the former need to survive in space.

Microbial Life in Extreme Environments, ed. D.J. Kushner, Academic Press, London, 1978, 412.

4
Life in space and its arrival on Earth

We diverge now in two major respects from the panspermia theory of Svante Arrhenius. In our view the arrival at the Earth of living cells, and of fragments of genetic material more generally, is a continuing ongoing process that directs the main features of biological evolution. It is this process which does the job that is usually attributed to Darwinism. For Arrhenius, on the other hand, it was the arrival of living cells early in the Earth's history which chiefly mattered. Thus Arrhenius concludes *Worlds in the Making* with the pessimistic paragraph:

> There is little probability, though, of our ever being able to demonstrate the correctness of this view by an examination of seeds falling down upon our Earth. For the number of germs which reach us from other worlds will be extremely limited—not more, perhaps, than a few within a year all over the Earth's surface; and those, moreover, will presumably strongly resemble the single-celled spores with which the winds play in our atmosphere. It would be difficult, if not impossible, to prove the celestial origin of any such germs if they should be found . . . .

In this chapter we shall find that the 'germs' instead of being only 'a few within a year' may well be of the order of 10^{20} within the year, and that there is evidence of their arrival to be found on all hands.

The reason for the pessimism of Arrhenius lay in the inefficiency of his method for dispersing cells from one star system to another. This was supposed to occur directly from a planet in one system to a planet in another system, and there are then serious difficulties at both ends of the journey. At launch, as it were, the cells were thought to be carried aloft in the atmosphere of the parent planet by convection currents, but as Arrhenius himself argues:

> ... air currents can never push spores outside our atmosphere. In order to raise them to still higher levels we must have recourse to other forces, and we know that electrical forces can help us out of almost any difficulty. At heights of 100 km the phenomena of the radiating aurora take place. We believe that the aurorae are produced by the discharge of large quantities of negatively charged dust coming from the Sun. If, therefore, the spore in question should take up negative electricity from the solar dust during an electric discharge, it may then be driven out (into space) by the repulsive charges of the other particles.

This passage is a fine example of ingenious thinking. Without the advantage of contemplating the wizardry of modern electronic systems, Arrhenius had already concluded that electrical forces could help him out of almost any difficulty!

In space, living cells can ride on the pressure of light from the Sun and other stars. As we saw in the preceding chapter, on this question the arguments of Arrhenius were entirely correct. But on the manner of arrival of cells from the solar system at other star systems his further argument contained defects. It was argued that, on approaching another star, radiation pressure from that star could check the speed of a travelling cell. If the speed were checked at just the distance from the star at which a planet happened to be, a gentle encounter with the atmosphere of the planet could ensue, the cell surviving entry into the atmosphere without becoming unduly heated.

It was at this point that the panspermia theory ran decisively off the rails, for this sequence of events would be highly coincidental and so could happen only rarely. This was exactly why Arrhenius considered that the number of 'germs' arriving at the Earth's surface from other star systems would be limited to only 'a few within a year'. So it came about that the possibility of understanding a wide range of biological phenomena, some of which we shall consider later in this chapter, was missed.

The trouble came because Arrhenius accepted the conventional view of astronomers (of whom ironically he had a poor opinion) that interstellar space was essentially empty. Even clear-cut examples to the contrary, like the well-known gaseous nebula in the 'sword' of the constellation of Orion, were regarded as inconsequential irrelevancies. Nowadays we know there is gas everywhere in interstellar space, sometimes in clouds like the Orion nebula and sometimes in a much

more diffuse form with a density of about one thousand atoms per litre—
the density in the Orion nebula ranges from about a million to one
thousand million atoms per litre. The crucial point is that biomaterial
ejected from a star system would not move through interstellar space
until it encountered and was retained by another star system. Such biotic
material would move through interstellar space until it encountered a
cloud of gas, and it would be retained by that cloud. The appropriate
logical scheme for the spread of biotic material, which is to say protozoa,
bacteria, viruses, or genetic fragments such as viroids, is therefore the
cycle of Figure 4.1, not the direct transfer from star to star as postulated
by Arrhenius. Direct transfer would be enormously weaker than the
effects of Figure 4.1, weaker perhaps by a factor 10^{20}. Let us see next how
this enormous jump in importance arises.

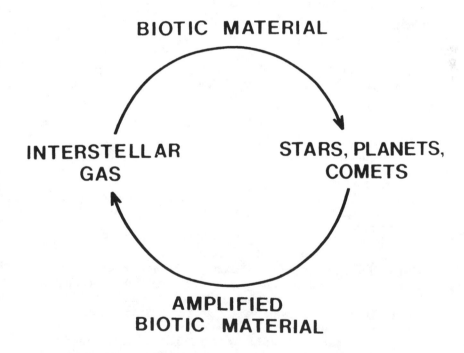

Fig. 4.1 Amplifying cycle for cosmic biology

There has been a possible circuit of the loop of Figure 4.1 for every
star that has condensed from the interstellar gas, of which there have
been about 10^{11} in our galaxy. Each loop offers the opportunity for

contributing a measure of biochemical evolution to life in the galaxy. At this stage we shall omit, for reasons discussed in chapters 1 and 2, and because we shall return to it later, the enormous problem of how life began in the first place. Supposing it to be present, each new star system receives at its birth a store of genetic information, providing for the construction of crucial biomolecules like the enzymes, histones and t-RNAs. If life takes root and flourishes in the new system, there will inevitably be mutational changes, and these changes are returned to the interstellar gas along the bottom arm of Figure 4.1. Thus the loop, with its 10^{11} circuits, has great evolutionary potential. We suspect that evolutionary effects involving t-RNA molecules,[1] arose in the loop of Fig. 4.1 and not at all on the Earth, as is usually supposed.

Let us see in a little more detail how the loop operated in a particular case, the solar system. In the earliest days of our system the primitive Sun was superluminous, so much so that the temperature at the Earth's distance from the Sun was about 1200°C, like a blast furnace. At the distance of the outer planets Uranus and Neptune, however, the temperature measured on the absolute scale (Kelvin) was about five times lower, about 300K, i.e. about 30°C, ideal for life processes to operate. Given a vast mass of chemical foods, given just the right temperature, and given the initial presence of the seeds of life, the number of living cells must literally have increased to an astronomical level, perhaps to as many as 10^{40} such cells. The situation is reminiscent of the story which Ibn Khallikán relates in the Life of Abú Bakr al-Súli:

> I have met many people who believe al-Súli to have been the inventor of chess, but this is an erroneous opinion, that game having been imagined by Sissa the son of Dáhir, an Indian, for the amusement of King Shihrám.
>
> It is said that, when Sissa invented the game of chess and presented it to Shihrám, the King was struck with admiration and filled with joy; he had chess-boards placed in the temples, and he expressed his opinion that the game was the best thing man could study, inasmuch as it served as an introduction to the art of war, and that it was an honour to the Faith and the World, as well as the foundation of all justice.
>
> The King also manifested his gratitude and satisfaction for the favour which Heaven had granted him in shedding lustre on his reign by such an invention, and he said to Sissa, 'Ask for whatever you desire'—'I then demand', replied Sissa, 'that a grain of wheat be placed in the first square of the chess-board, two in the second, and

[1] R.J. Cedergren, B. LaRue, D. Sankoff, G. Lapalme and H. Grosjean, *Proc. Nat. Acad. Sci.*, vol. 77, 1980, 2791.

that the numbers of grains be progressively doubled till the last square is reached: whatever this quantity of wheat may be, I ask that you bestow it on me.' The King, who intended to make him a worthy present, exclaimed that such a recompense would be too little, and he reproached Sissa for asking so inadequate a reward. Sissa declared that he desired no other gift and, heedless of the King's remonstrances, he persisted in his request. The King at last consented, and gave orders that the required quantity of grain be given to him. When the chiefs of the royal house received their orders they calculated the amount, and answered that they did not possess near as much wheat as was required. When these words were reported to the King, he was unable to credit them and had the chiefs brought before him; when questioned on the subject they declared that all the wheat in the world would be insufficient to make up that quantity. They were asked to prove the truth of this contention, and by a series of multiplications and reckonings they demonstrated that such was indeed the case. The King then said to Sissa, 'Your ingenuity in imagining such a request is even more admirable than your talent in inventing the game of chess'.

The way in which the doubling of the grains of wheat is to be done consists in the calculator placing one grain in the first square, two in the second, four in the third, eight in the fourth, and so on, until he comes to the last square, placing in each square double the number contained in the preceding one. I was doubtful of the contention that the final amount could be as great as was said, but when I met one of the accountants employed at Alexandria, I received from him a demonstration that convinced me that their statement was true: he placed before me a sheet of paper, on which he had calculated the amount up to the sixteenth square, obtaining the result 32,768. 'Now,' said he, 'let us consider this number of grains to be the content of a pint measure, and this I know by experiment to be true'—these are the accountant's words, so let him bear the responsibility—'then let the pint be doubled in the seventeenth square, and so on progressively. In the twentieth square it will become a *wayba*, the *wayba* will then become an *ardeb*, and in the fortieth square we have 174,762 *ardebs*; let us consider this to be the contents of a corn-store, and no corn-store contains more than that; then, in the fiftieth square we shall have the contents of 1,024 stores; suppose these to be situated in one city—and no city can have more than that number of granaries or even so many—we shall then find that the number required for the sixty-fourth and last square corresponds to the contents of 16,384 cities; but you know that there is not in the whole world a greater number of cities than that. This demonstration is decisive and indubitable.

This is just the way it is in biology, except that to yield 10^{40} grains of wheat, equalling the number of living cells that could have been produced in the early days of the solar system, the chess-board would need to have had 133 squares. Living cells multiply typically in generation times of two or three hours under terrestrial conditions. 133 such generations would require a mere couple of weeks. Multiplication times at the outside of the solar system, in the multitude of comet-like bodies that existed there initially, before Uranus and Neptune were formed, would be longer than this, of course, but there is no reason why the main amplification in number should not have occurred in a few orbital periods around the Sun. A few centuries, or a few thousand years, seems the likely time-scale in which the biological explosion would take place.

The potentiality of living systems to increase enormously in their numbers is never given real scope to operate in orthodox biology. The potential appears in the Darwinian theory, but only in a minor degree. Darwin himself wrote as follows in one of his notebooks:

> On the average every species must have some number killed year with year by hawk, by cold, etc.,—even one species of hawk decreasing in number must affect instantaneously all the rest. The final effect of all this wedging must be to sort out proper structure. . . . One might say there is a force like a hundred thousand wedges trying to force every kind of adapted structure into the gaps in the economy of nature, or rather forming gaps by thrusting out weaker ones.

The concept here is of a jostling for advantage. Small gaps appear and are immediately filled by an adapted animal through its potentiality for rapid expansion. Such effects, however, are only what mathematicians call 'small perturbations'. There is nothing of an explosion into a vast area of virgin territory, such as opens up in Figure 4.1 when a new star system is formed.

Amplified biotic material is returned to the interstellar gas in two ways, one quick and soon done with, the other maintained over a very long time-scale. It has been estimated that our solar system returned as much as one to ten per cent of the mass of the Sun itself to the interstellar gas. Most of the returned material would be hydrogen and helium, but according to our point of view perhaps one part in a hundred of the returned material could be biotic. From 10^{11} circuits of Figure 4.1 this return process would yield a total of biomaterial amounting to ten million times the mass of the Sun, a fantastic quantity compared to the amounts contemplated in orthodox biology (about 10^{21} times more!).

In addition to this quick return, in each star system biomaterial would be stored away inside comets. The comets of our own solar system have themselves been stored for more than four thousand million years on its outside, considerably beyond the distant planet Neptune. In the great cold which now exists there, the biomaterial would be hard-frozen, with life-processes entirely suspended. From time to time, however, a comet may be deflected by the gravitational field of a passing star in such a way that it follows a new orbit that comes to the inner part of the solar system. For a particular comet this is an improbable happening, but there are so very many comets that it happens every year for two or three of them, as it has happened for Halley's comet, whose present orbit is shown in Figure 4.2.

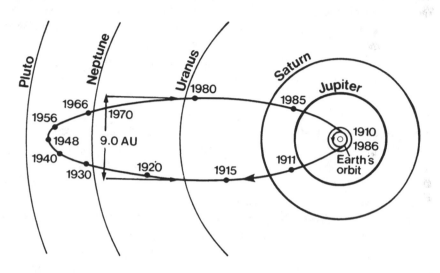

Figure 4.2   Orbit of Halley's comet

In such a case as Halley's comet, the frozen material tends to thaw at its surface during its passage through the inner regions closest to the Sun. Evaporation of gas and particles produces the visible coma and tail of the comet. Radiation pressure now expels particles of appropriate sizes at high speeds from the solar system, back to the clouds of gas that lie between the stars. The cells produced in the 'biological bomb' which occurred in the early days of the solar system are thus ejected out into space, again completing the loop of Figure 4.1, this time in a slow, persistent way.

The ability of living cells to withstand large X-ray doses can now be assessed in its proper setting. Living cells travelling in space, even if not close to one particular star, are still exposed to X-rays and to cosmic rays (which act on them rather like radioactive particles). For living material, with its large chemical content of carbon, nitrogen and oxygen, X-rays are considerably more damaging than cosmic rays. X-rays are contributed in part by a large number of ordinary stars like the Sun, in part by a much smaller number of specially intense X-ray stars, and in part by other galaxies, especially galaxies that are exceptionally strong X-ray sources.

For an unshielded cell at a typical point of interstellar space the most deleterious effects come from soft X-rays. Calculation shows that a dose of a million rads due to soft X-rays would accumulate in about ten thousand years, which is about the time a cell moving at a speed of 100 km per second would need to cover the distance of three light years by which neighbouring stars are typically separated. For cells inside a cloud like the Orion nebula, however, there would be shielding by the gas of the cloud against soft X-rays. Harder X-rays would still penetrate the cloud, but would need a million years to irradiate cells with a dose of a million rads. Thus inside the interstellar clouds, living cells with the known ability to withstand radiation doses of the order of a million rads would have quite long lifetimes. Indeed the lifetimes could well be considerably longer than these estimates might superficially suggest, because X-ray doses accumulate only very slowly in space, away from the vicinities of stars, whereas lethal doses measured for cells in the laboratory are for sudden flashes—and experience shows that sudden flashes at high intensity are far more damaging than slowly accumulated doses which permit ample time for repair processes to operate.

It is interesting to consider the conditions under which living cells could make the journey, not merely from star to star, but from one galaxy to another. The speeds with which particles of appropriate sizes in the tails of comets are expelled from the solar system are generally about 100 km per second, but higher speeds are certainly attainable. Such higher speeds arise, for instance, if particles ball up into a loose structure with an effective density perhaps only one-tenth of water, a situation which is actually believed to happen for some cometary particles. In sufficiently exceptional cases, speeds generated by radiation pressure of 1000 km per second might well arise, and a cell travelling at this speed could make the journey from our galaxy to the neighbouring large galaxy in the constellation Andromeda in less than a billion years, in about one-fifth of the span of time for which life has existed here on the Earth.

It is easy to see from the example of Halley's comet in Figure 4.2 why the present point of view differs so markedly from the panspermia theory of Svante Arrhenius. Comets shed material most effectively when they are closest to the Sun, which for Halley's comet is in the region of the Earth, and also in the region of the Earth for the several other comets which come each year to the inner parts of the solar system. The Earth is therefore perpetually situated in a halo of particles shed from comets. It has been estimated that the total mass of such particles entering the Earth's atmosphere is about 10^{10} grams per year, sufficient if they were all bacteria to give about 10^{22} bacteria incident on the Earth each year, or if they were all virus particles to give about 10^{24} such particles. Even if we allow generously for loss of viability during the journey from comet interiors to the surface of the Earth, the amount remains very large, enormous compared to Arrhenius's estimate of only 'a few within a year'.

If the entry speed into the terrestrial atmosphere is too high, viability may be lost through flash heating, due to the sudden frictional resistance of the atmospheric gases. In *Diseases from Space* we made calculations which showed that at typical entry speeds (40 km per second) there is a maximum diameter of about one micrometre up to which living cells can remain viable with respect to flash heating. This maximum diameter admits viroids, viruses, most bacteria, but not eukaryotic cells—protozoa and microfungi. However, the atmospheric entry speeds can in exceptional cases, for particles in favourable orbits, be considerably less than typical speeds. The maximum permitted diameter then rises to about sixty micrometres, which easily admits eukaryotic cells. Thus atmospheric heating is no impediment to living cells reaching the surface of the Earth, although a considerable fraction of the larger cells would be destroyed by heat.

While entry into the Earth's atmosphere thus presents a hazard, safe entry would be impossible without an atmosphere. The cells would then make hard landings at high speeds in excess of 10 km per second. They would crash into a solid surface, as in fact small particles do at the surface of the Moon, and would be totally destroyed by the extreme violence of their impact.

'Soft' landings would occur in the atmospheres of other planets as well as the Earth. There would, in particular, be landings of viable cells on the planet Mars. There are three possibilities for such cells. They may take root and multiply as on the Earth. They may find the conditions too inhospitable and die, or they may find the conditions suitable for remaining in a resting state but not suitable for them to multiply. We consider the last of these possibilities to be most likely on Mars. We expect living cells, bacteria especially, to be present there but not in the enormous numbers that have become established by replication here on the Earth. The natural question is whether the *Viking 1* and *Viking 2* landers on Mars would have detected a thin surface distribution of bacteria in their resting states. We doubt they would. Indeed we doubt that similar landings at two inhospitable places on the Antarctic plateau would have revealed the profusion of microbial life which actually exists in Antarctica.[2]

Sending out space vehicles to detect living cells in the neighbourhood of the Earth is by no means a straightforward matter, since the collisional velocities between the cells and the vehicles would typically be about 40 km per second, and the cells would instantly be destroyed on impact, as at the surface of the Moon. For a physical collection process to be meaningful in a biological sense it is essential that the encounter speeds be cut almost to zero. The sensible thing therefore is to use soft landings in the Earth's atmosphere to slow the cells, and then to collect them with equipment attached to a balloon.

This raises, however, the problem of eliminating terrestrial contamination, both contamination on the equipment itself and by cells from the ground being carried upward into the high atmosphere by air movements. The first of these difficulties can be resolved by attention to the design of the experiment and can be tested for by also flying suitable control packages. The second difficulty can be eliminated, one would think, by balloons going well above the tropopause, say to heights of order 30 km or more (about 90,000 feet). The atmospheric temperature rises *upwards* from about -55°C at the tropopause (height about 12 km) to about 0°C at the stratopause (height about 50 km). Such an inversion of the temperature gradient with respect to the direction of gravity strongly inhibits all local vertical air movements, although slow upward and downward interchanges of air affecting the whole atmosphere are not excluded. It is generally conceded, however, that such slow interchanges are less important than gravity for particles with the sizes of bacteria and microfungi. Thus any such particles should have vertical motions that are predominantly downward. Indeed for living cells of such sizes we should expect to find few, if any, lifted upward from below the tropopause. Living cells midway between the tropopause and the stratopause, at a height, say, of 30 km (about 90,000 feet), should be from space rather than from the ground.

A test for a space origin can also be made. Particles with a size of half a micrometre, typical for bacteria, fall through still air about four or five times faster in the height range from 90,000 feet down to 60,000 feet than they do in the height range from 60,000 feet down to 30,000 feet. This slowing of the rate of descent arises because of increased air resistance at the lower heights. Since particles incident from space at a steady rate attain a smooth downward flow, it follows that such particles

---

[2] For a detailed discussion of this question see R.E. Cameron, R.C. Honour and F.A. Morelli in *Extreme Environments*, ed. M.R. Heinrich, Academic Press, New York, 1976.

would need to have a density that behaved reciprocally with respect to their downward speed, four to five times greater density for the height range from 30,000 to 60,000 feet than for the range from 60,000 to 90,000 feet. This was exactly the result reported by V.W. Greene, P.D. Pederson, D.A. Lundgren and C.A. Hagberg for a series of balloon flights between 1 August 1962 and 1 October 1965.[3] These workers found living cells in the high atmosphere, bacteria mostly, with a density of one per ~ 400 cubic feet for the height range from 30,000 to 60,000 feet, and with a density of one per $\sim 2,000$ cubic feet for the height range from 60,000 to 90,000 feet. Using a downward speed of 3.10^{-3} metres per second for the lower of these ranges, and taking the cells to be incident from space, it is easy to calculate that the number of viable cells with sizes upwards of half a micrometre that reach the surface of the Earth from space would be about 5.10^{18} per year, a far cry indeed from Arrhenius's 'a few within a year'.

To deny this conclusion it is necessary to suppose that large-scale air movements carry many more bacteria from low levels to great heights in the atmosphere than the strong temperature inversion between the tropopause and the stratopause would appear to permit. If there were a non-steady incidence of cells from space, however, even this *ad hoc* supposition could be dismissed. Thus a sudden large influx of cells would produce a situation for a while in which the density of cells in the high atmosphere actually increased upwards, a situation it would surely not be possible to attain by mass air movements lifting cells from below.

Three balloon missions were flown during 1963 to altitudes of 125,000-130,000 feet, by G.A. Soffen at the Jet Propulsion Laboratory (JPL).[3] There was a malfunction on the second flight. In the first flight seven *Penicillium* bacteria were found from a sample volume of 1000 cubic feet, and in the third flight five such bacteria were found. Compared to the results of Greene, Pederson, Lundgren and Hagberg these JPL flights indicated a considerable increase with height in the bacterial density, a conclusion found so surprising that the experiment was thought to be flawed by contamination and was discontinued.

These balloon results are now more than fifteen years old, and it might be thought that by 1980 the matter would have been settled beyond all argument over issues of contamination. We ourselves searched the subsequent literature in some excitement at what we would find there, but to our great surprise we found—nothing! The situation was so curious that we made special inquiries as to what had happened,

[3] NASA report N65-23980.

finding to our even greater surprise that the experiments had been discontinued because NASA had withdrawn the supporting funds.

5.10^{18} per year is a lot of bacteria, and with the possibility of the incidence of viroids and viruses being two orders of magnitude larger still, $\sim 10^{21}$ per year, it was clear that these amounts should easily be detectable, if a fraction of them were pathogenic. It was easy to make an interesting calculation along these lines.

If the bacteria were to fall in still air to the ground, their downward speeds would slow more and more because of the increasing air resistance at the lower levels, and their density would rise reciprocally as before, from one bacterium in ~ 400 cubic feet for the altitude range from 30,000 to 60,000 feet to one bacterium in ~ 100 cubic feet at ground level. People take in about 250 cubic centimetres of air at each breath, and they normally breathe about twenty times per minute—which is to say about ten million breaths per year. Thus people breathe about 2500 cubic metres of air in a year, which is nearly 100,000 cubic feet. Hence at a density of one bacterium per 100 cubic feet, people would breathe about a thousand space-incident bacteria per year, and they might well breathe of the order of a hundred thousand viroids and virus particles, quite sufficient to produce devastating pathogenic effects.

It was these considerations that led us eventually to our book *Diseases from Space*. Pathogenic attacks, even such minor afflictions as the common cold, are obvious to the eye. So here was another of Nature's signposts, a signpost so clear-cut that it needed no large funds from any agency to read the lettering on it.

Motions of the lower air do not change the average situation much, but they tend to produce vagaries, with the bacteria and viruses falling in a patchy distribution to ground-level. Rain and snow can have a considerable effect, however, by cleaning the bacteria and viruses out from the lower air. We would then tend to swallow them instead of breathing them, as is easily seen from another simple calculation. People take in about half a gallon of liquid a day. Before the boisterous advent of modern civilization with its beers, pepsis and cokes, most of the liquid which people drank was plain water, and most of the water was surface run-off, recently fallen rain or snow.

Now half a gallon of water a day adds up to about five-sixths of a cubic metre of water per year. Since the annual precipitation averaged over the Earth is equivalent to a depth of ~ 0.8 metres of liquid water, it follows that under natural conditions people consumed the water falling as rain or snow on about a square metre of ground. 5.10^{18} bacteria falling to the surface per year gives about 10,000 bacteria incident on each square metre of ground. It follows therefore that if the bacteria fell in rain or snow, and if they remained in the surface water, then people would drink in each year the water containing ten thousand bacteria. Of course not all the bacteria would remain in the surface water. Some would be taken up in the soil. Some would adhere to plants and trees. Even so, it is likely that people would drink at least a thousand space-incident bacteria per year and perhaps as many as 100,000 viroids and virus particles.

The effect of rain and snow would thus be to switch the intake of viroids, viruses and bacteria from the respiratory tract to the stomach and gut. Since the respiratory tract is on the whole an area that is more sensitive to pathogenic attack than the stomach and gut it was necessary we suspect that animals developed the ability to separate the air-channel going from the nose to the lungs from the liquid-channel going from the mouth to the stomach. In *Diseases from Space* we speculated that a snout was a necessary part of this development for animals living in the open, away from the shelter of large trees:

> The human nose is not a directly scaled-down version of an animal snout, such as is possessed by dogs, pigs, bears, cattle and so on—animals that have evolved out in the open. The human nose is a built-up version of an ape's nose, which consists externally of little beyond two adjacent holes in the face. An ape's nose is probably fine for a creature living in the forest, protected from infalling pathogens by a thick canopy of leaves overhead. But an ape's nose would be a poor thing for a creature that came out of the forest into open ground, since it would hardly be possible with such a nose to avoid sniffing up rain drops. Equipped only with two holes in the face, the incidence of pathogens could be an order of magnitude greater than it is with the human nose. An ape coming from the forest into the open therefore had to evolve a beak of a nose, from the end of which rain drops would drip rather than be sniffed up into the nasal passages.

There is actually a monkey with a substantial nose, but it lives in a watery, swampy environment that is comparatively free from high tree cover. This is the proboscis monkey of Borneo. To continue the above quotation:

> It is very curious how much emotional importance we attach to the shape of the human nose, and how strongly significant it is in our assessment of a person's looks.

A monkey's looks too, according to the pictures one sees of the proboscis monkey. It is likely that mate-selection in accordance with such responses is at least as important a means of sieving the gene pool, at any rate among the higher animals, as are the more usually discussed muscular tests for survival.

Although 5.10^{18} must be considered large for the annual incidence of bacteria from space, it is still very small compared with the total number of bacteria resident at the Earth's surface, which is probably in the range 10^{27} to 10^{28}. However, a space-incident bacterium can increase in number extremely quickly if it finds the conditions suitable. The resident population would therefore be expected to be large compared with the annual incidence rate.

It is a mistake to concentrate too much on pathogenic bacteria, which comprise only a tiny fraction of the total. Nor should we concentrate on photosynthetic bacteria, which also form only a fraction of the total. Like businessmen, bacteria are to be found wherever a profit is to be made, an energy profit. They feed on exothermic chemical reactions that go only very slowly under inorganic conditions, by greatly speeding them up. Put carbon dioxide and free hydrogen together under a moderate pressure and they can be stored without change essentially for an eternity. But allow methanogenic bacteria into the mixture and it will soon be changed into methane and water, with a small energy-profit going to the bacteria for their services.

Bacteria and microfungi are to be found more or less everywhere, in hay ricks, oil wells, birds' nests, mine tailings, coal dumps, in our stomachs, and in enormous profusion in the soil. Every bit of reasonable arable land has about a third of a pound weight of bacteria to each square metre of ground, the weight of a child's bag of sweets. Something like a half of the dry weight of farmyard muck is made up of bacteria. So when we buy a load of muck to fertilize our gardens about half the money goes for bacteria.

The money is well-spent, for without bacteria the photosynthetic plants would convert all the carbon dioxide in the Earth's atmosphere into solid materials, mainly cellulose and other polysaccharides like starch. This would happen in only a few years, the principal life-forms of the Earth would then die, and over the following aeons the polysaccharides would gradually become degraded into coal. It is bacteria that save us from this depressing scenario, by breaking down the material of dead plants in such a way that the carbon dioxide is returned to the atmosphere, and so that nitrogenous chemicals which would otherwise be locked away in the dead material are made free in the soil and so available to other living plants.

It is the same story within the bodies of animals. We always talk as if we ourselves digested our food. This is loose talk, for it is bacteria that break the food down for us into more elementary substances which our bodies are able to use. Bacteria do much of the digesting. We only create the conditions that make it convenient for them to live inside us.

Pathogenic bacteria in the gut are hard to treat medically except with a laxative, since antibiotic drugs hit the essential bacteria as well as the pathogenic ones. By eating too much of some things and too little of others, some health-food addicts also tend to create digestive problems for themselves. It is as well, where diet is concerned, to remember that an Indian without digestive bacteria is a dead Indian—and the white man too.

The ubiquity of bacteria suggests that they are highly adapted to their environment. In the sense that bacteria can survive and prosper in the most rugged environments, this is true. An outstanding case of bacterial toughness was reported some years ago by H. Dombrowski, who succeeded in isolating and cultivating bacteria from salt deposits that were laid down some two hundred and fifty million years ago, in the so-called Zechstein Sea which then covered much of N. Europe.[4] Dombrowksi concludes the summary of his paper with the statement:

> Studies on other salt deposits were made, and living bacteria were isolated from salt deposits from the Middle Devonian, the Silurian, and the Precambrian.

Thus bacteria managed to survive inside the Earth for six hundred million years or more.

Bacteria, however, are not well-adapted to their environments in a Darwinian sense. Reference back to the earlier quotation from Darwin's notebook shows his concept to be one in which each species managed to survive in fierce competition with others by maximizing its advantages. Bacteria can be subject to situations in which this concept becomes valid, as for instance in hospitals where pathogenic bacteria are subject to strong selection for resistance to particular drugs. In Nature, however, the situation is quite otherwise. Different species of bacteria mostly exist with little interaction, and they are often found in conditions far removed from optimum.

The facts exactly fit the picture of an exceedingly hardy, all-purpose collection of cells incident from space that simply takes root and establishes itself on the Earth almost everywhere, optimum conditions or not. Thus a considerable fraction of the bacteria found in warm temperate soils are of the so-called psychrophilic kind, which is to say they would reproduce themselves better at temperatures lower than those which exist in such soils—conditions for psychrophilic bacteria tend to be optimum at about 15°C.

[4] H. Dombrowski, *Annals New York Academy of Sciences*, vol. 108 (Pt 2), 1963, 453.

Thermophilic bacteria, with optimum temperatures of about 55°C, are found in arctic soils and waters, the opposite situation. Thus R.H. Mcbee and V.H. Mcbee obtained thermophilic bacteria in samples of soil and ocean water collected near Point Barrow in N. Alaska.[5] In their paper these workers remark:

> There is no doubt that thermophilic bacteria are present in Arctic soils and water, and that they may constitute an appreciable portion of the bacterial population in some specimens. Their presence is not correlated with finding of coliform bacteria, so it appears that it is not due to fecal contamination by birds and other animals. . . .
>
> Insufficient work has been done to develop an explanation of how and why thermophilic bacteria occur in regions where the soil is frozen for most of the year and seldom reaches a temperature of 10°C, even at the surface. The fact that thermophilic bacteria exist under such conditions and have also been found in deep ocean bottom cores where the *in situ* temperature was below 10° indicates that the physiology of these organisms needs further study.

Life is a long-term affair in the deep ocean, where the temperature is actually about -1°C, with bacteria having plenty of time to adapt to the high pressure of the overlying water. Yet in defiance of the lightning-quick response which bacteria are supposed to have in pathogenic environments there has apparently been little adaptation in the deep ocean. Thus R.E. Marquis and P. Matsumura write:[6]

> Of necessity, this article will differ in basic outlook from many of the others in this book, largely because we are as yet unsure that any organisms have undergone specific adaptations to life under pressure. That some bacteria are highly baroduric seems at the present time to be purely fortuitous and not related to specific adaptation. In fact, Kriss was able to extract baroduric bacteria from garden soil.

So a 'fortuitous' ability (i.e. not subject to a Darwinian explanation) to withstand great pressure has to be added to a fortuitous ability of bacteria to withstand enormous doses of X-rays and the fortuitous ability to withstand very low pressures also, as well as temperatures close to absolute zero.

There are also signposts from the biochemistry of bacteria to show that their origin was not terrestrial. Photosynthetic plants convert carbon dioxide and water to sugars and polysaccharides. The use of

---

[5] *Journal of Bacteriology,* vol 71, 182.
[6] *Microbial Life in Extreme Environments,* ed. D.J. Kushner, Academic Press, New York, 1978, 109.

water would make sense in a terrestrially oriented theory because water is abundant in the biosphere of the Earth. Photosynthetic bacteria, on the other hand, use either free hydrogen or hydrogen sulphide in place of water, both rare materials on Earth. This substitution would be absurd for cells with a terrestrial origin.

Bacteria are found occupying tiny specialized niches. Thus J.G. Zeikus and R.S. Wolfe isolated a highly thermophylic methanogenic bacterium which required conditions for replication that were very peculiar, namely an atmosphere with a 4 to 1 mixture of free hydrogen to carbon dioxide and a temperature of at least $40°C$.[7] The optimum temperature for growth was $65°C$ and free oxygen had to be strictly absent. It may be wondered where on the Earth such conditions exist. The answer is in sewage sludge, a product of modern industrial society.

Such cases fit well with our picture. The cycle of Figure 4.1 contains bacteria with the ability to exist in a range of environments much wider than anything found on the Earth. Bacteria arrive here from space with the full range of cosmic properties, and terrestrial conditions simply filter out the restricted subset that can survive when they arrive at ground-level. This subset depends on available chemicals—rocks, soils, ocean water with its dissolved contents, mine tailings, sewage sludge, volcanic hot springs, birds nests, and so on. Whenever any new environment, however specialized or small it may be, arises either from natural or man-made causes, new bacteria from the wide spectrum of cosmic possibilities are available to take advantage of it.

Journal of Bacteriology, vol. 109, 1972, 707.

5
Almost from its beginning the Earth had life

In this chapter and the next we discuss the fossil record of life on the Earth, shown in outline by the 'biological clock' of Figure 6.1. Here we shall be concerned with single-celled forms, while in chapter 6 we consider multi-celled forms, the so-called metazoa. From Figure 6.1 it will be seen that whereas the history of the metazoa has been confined to the last 700-800 million years, the single-celled forms have been traced backward in the fossil record to almost the first moment when it was physically possible for life to exist on the Earth. This was about 3.8 billion years ago.

There are two kinds of single-celled forms, distinguished by size and complexity of structure. The smaller and simpler prokaryotic cells are represented by bacteria and the blue-green algae. They are characteristically about 1 micrometre in size and their DNA is diffusely distributed. The larger eukaryotic cells have dimensions typically of the order of 10 micrometres and their DNA is concentrated in a nuclear region, or sometimes in several such regions. The distinction between these two types is shown in Figure 5.1, which compares drawings of the bacterium *E. coli* (prokaryote) and of a yeast cell (eukaryote).

Although prokaryotes may cluster together to form colonies, as for instance the colonies of blue-green algae known as stromatolites, prokaryotes do not form interconnected aggregates of cells in the manner of eukaryotes. Thus multi-celled plants and animals, the metazoa, are made up of eukaryotic cells.

Eukaryotes occur in single-celled forms as algae (excluding the blue-greens), as microfungi, and, according to traditional biological classification, as protozoa. In variety and number—there are about 30,000 living species—the protozoa are almost as omnipresent as bacteria. The most widely scattered are the marine plankton which provide bases for the food chains of oceanic life. Many species are parasitic on multi-celled organizations, and for this reason cannot be regarded as primary life-forms. Some protozoa are so large (millimetres or even centimetres in size) that it is doubtful whether they should be thought of as single-celled, and indeed many protozoologists prefer nowadays to avoid this description for the whole phylum of the protozoa.

Algae are photosynthetic, using water, carbon dioxide and sunlight to produce sugars and polysaccharides such as starch and cellulose. The molecule of chlorophyll plays a crucial role in this process. Microfungi do not contain chlorophyll, however, and so microfungi must obtain their sugars and polysaccharides from sources outside themselves. Photosynthesis, or the lack of it, was used traditionally in biology as a means to distinguish plants from animals, the algae being plants and the microfungi being thought of as animals. The distinction has come to seem less useful, however, as it has become apparent that some microfungi, the slime moulds for example, exhibit properties of both animals and plants. Like the protozoa, both algae and microfungi are widely distributed over the Earth. Microfungi have some 50,000 living species, even more than the protozoa.

In the past it seemed natural to most biologists to suppose that eukaryotes evolved from prokaryotes. Both kinds of cell make use of closely similar enzymes. Indeed, enzymes from prokaryotes (*E. coli,* for example) can be used to promote biochemical processes inside eukaryotes, as for instance the transcription of DNA into RNA. Prokaryotes were regarded as primary, partly because they seemed simpler in structure than the eukaryotes (Figure 5.1) and partly because their genetic content is considerably less. Thus a yeast cell has sufficient DNA to provide for about ten times more genes than the DNA of a typical bacterium.

If one believes that life originated on the Earth, the compulsion to search for an ancestral cell is strong, and the tendency is to imagine that there must have been a time when simple cells existed, but when complex cells did not. Prokaryotes were therefore believed to have had a more ancient history than eukaryotes. Bacteria and blue-green algae are known to have existed for at least 3,000 million years, but eukaryotic cells were thought not to have emerged until 1,000 to 1,500 million years ago. As we shall see below, this belief has turned out to be wrong.

Nor do recent discoveries in microbiology support the idea that

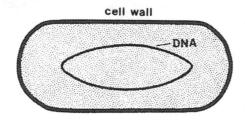

Fig. 5.1 **(a)** Drawing of the bacterium *Escherichia coli*

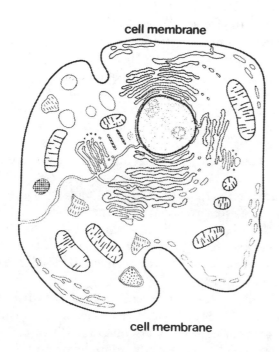

Fig. 5.1 **(b)** Drawing of a yeast cell

eukaryotes evolved from prokaryotes. The genes of the latter are continuous sequences of bases on the DNA, whereas the genes of eukaryotes are built characteristically from several disjointed segments of DNA. There is evidently a major chasm between the modes of gene expression in the two kinds of cell. A similar conclusion might have been reached long ago from the fact that photosynthesis in prokaryotes does not use water as in eukaryotes, a remarkable difference mentioned already in chapter 4.

We turn now to the fossil record. Fossilization is a general term applied to any process by which the remains or traces of plants or animals are preserved in the Earth's crust. Usually, but not always, for an organism to be preserved two conditions must be satisfied: the organism must possess hard parts such as bones or a shell or a cell wall that are capable of being fossilized, and there must be rapid burial of the remains by inorganic material in order to prevent decomposition and to exclude scavengers.

There are, however, exceptional cases that break these general rules. Several million years ago a hominoid walked in soft mud at the edge of a lake, leaving a trail of footprints. It happened that the lake dried up and that the mud hardened to give a firm cast to the prints. Then the winds blew dust over the area, accumulating sufficiently on top of the prints to protect them from erosion down the ages, until man came along to dig through the protective layer and so to reveal the prints of his distant forebear.

But exceptional cases like this are not relevant when very long geological time-scales are involved, as we are involved here in seeking to penetrate back in time more than 2,500 million years, back to the oldest Archean period of geology. Nor can we expect biological structures to be preserved literally, as they are when actual bones are found in the case of more recent fossils.

Most fossils are altered from their original biological composition by the substitution of more preservative inorganic minerals. This can happen in a way that maintains the original characteristic biological shape, as for instance the shell of a marine organism, or the shape of a colony of algae or bacteria. Common replacement materials are calcium carbonate, hematite and silica.

Favourable conditions for such replacements occur when sedimentary rocks are being laid down at the margins of continents. Two opposing processes are constantly at work on the Earth's crust. Motions of the oceanic 'plates' tend to compress the continents into smaller areas, causing ranges of high mountains to be formed. Erosion of all high ground by wind and water denudes the mountains of material, which is then carried by rivers to the margins of the continents, where submerged oceanic shelves of 'sedimentary' material are gradually built up. Thus plate motions compress the continents while erosion expands them. Marine organisms and also land-based organisms carried by river waters tend to be buried and fossilized in the sedimentation process at the continental margins, and so rocks formed by this process, the sedimentary rocks, tend to be rich in fossils. Examples are the chalk deposits of southern England and the limestone hills of northern England. The former are rich in the fossils of marine organisms that existed when the chalk was laid down about 100 million years ago, and the latter are rich in the fossils of organisms that existed when the limestone was laid down, about 350 million years ago.

Although most biochemical material is perishable on the very long time-scale of thousands of millions of years that is of interest in this chapter, there are some exceptions. The chlorophylls of green plants, algae and photosynthetic bacteria contain a side chain consisting of the substance phytol (phytyl alcohol) which in conditions of fossilization loses a water molecule to become the hydrocarbon phytane, which is then hard to destroy. Thus a suspected microfossil with most of the original biochemicals now replaced by inorganic minerals may still contain phytane. Since phytane is not produced abiologically in the sedimentation process, a positive detection of its presence gives assurance that a suspected fossil must be genuine.

The carotenoids are a group of yellow, orange and red pigments which appear widely in plants (petals of flowers, citrus fruits, tomatoes, carrots). They are synthesized by bacteria, algae, protozoa and microfungi, and like phytane they are resistant to destruction. The detection of carotenoids within a microfossil is also a guarantee of its genuineness, because they too are not produced abiologically in the sedimentation process.

Up to the middle of 1979 the earliest geological evidence for microbial life was dated at about 3,200 million years ago. This evidence was found in sedimentary rocks from the Swaziland Supergroup in the Fig Tree and Onverwacht formations. Clusters of microfossils in the form of bacillus-like rods and algae-like spheroids were found. The sizes, shapes and inner structures of these clusters in the rocks, as seen for instance in electron micrographs, point to their being prokaryotic organisms. Furthermore, the presence of phytane shows them to have been photosynthetic prokaryotes.

In 1980, several workers reported the presence of bacterial fossils in the Archean rocks of Australia, dated at about 3,500 million years ago. If it could have been shown that eukaryotic cells were then absent, the positive detection of prokaryotic cells so early in the history of the Earth would accord with the view that eukaryotes have evolved from prokaryotes, and this was the sense in which many biologists have interpreted the evidence. In 1979, however, H.D. Pflug, H. Jaeschke-Boyer and E.L. Sattler reported structures in the Swartkoppie cherts, with an age of about 3,400 million years, that were remarkably similar in shape, size and formation to modern yeast cells.[1] If these structures were indeed fossilized yeast cells, the ancestry of eukaryotic cells has been pushed back very far in time, since until this work the oldest suspected fossilized eukaryotes dated from about 1,400 million years ago.

Because of the potentially revolutionary significance of the Swartkoppie microstructures, the investigators were careful not to claim them to be microfossils simply from their morphological similarity to yeast cells alone. Taking at first a neutral position they named them *Ramsaysphaera*.

In August 1979, Pflug and Jaeschke-Boyer also reported yeast-like structures in a metamorphosed rock from the Isua region of West Greenland, a region dated at about 3,800 million years ago, the world's oldest known rocks. Once again, the investigators took an initially neutral position, naming these further occlusions *Isuasphaera*. To proceed further with the work, it was necessary to determine whether or not organic materials of biological origin were present within the *Isuasphaera*. Professor Pflug explained to us in private conversation that he was unwilling to grind up the rocks containing the occlusions for chemical analysis, because however great the precautions that were taken against contamination from present-day sources of biochemicals it would never be possible to insist categorically that there had been no contamination. Therefore the *Isuasphaera* had to be examined *in situ,* within the silica matrix of the surrounding rock. Sections of the rock could be cut that would be consistent with this requirement (40 to 100 micrometres thick) and through which radiation probes could penetrate. The characteristic absorption bands of organic residues could then be looked for, provided the technical problem of focusing the probes *within* the *Isuasphaera* could be solved. In this, Professor Pflug was given valuable assistance by the German optical companies, with the result that organic residues were indeed found within the Isua microstructures.

As things now stand, therefore, eukaryotes appear to have been detected at an even earlier date than prokaryotes, although the difference between the age of 3,800 million years for the Greenland rocks and 3,500 million years for those from Australia is not very relevant. What is important is

[1] H.D. Pflug, H. Jaeschke-Boyer and E.L. Sattler, *Microscopia Acta,* vol. 82, 1979, 255.

that eukaryotes seem to be just as old as prokaryotes, and that the Isua and Swartkoppie microfossils are morphologically identical to modern yeasts.

The Earth itself is about 4,500 million years old. The bulk of the material that went to form our planet condensed from a hot gas expelled by the primitive Sun. The condensation process occurred at temperatures approaching those of a blast furnace. Thus primitive Earth conditions, 4,500 million years ago, were surely sterile to life. At the very beginning of its history, our planet must have presented a stark and arid prospect, not unlike the appearance of the present-day Moon.

Comets came to the rescue to change this barren landscape, not only by supplying living cells, but by providing the Earth with water and other volatile materials. This crucial endowment of life-supporting chemicals was made possible by the many billions of comets which surround the solar system, and by the gravitational effects explained in chapter 4, whereby comets are sometimes perturbed into orbits which bring them to the inner regions of the solar system, thus making encounters with the Earth inevitable for a small fraction of them.

Even today, comparatively minor collisions of denuded cometary nuclei (the so-called Apollo objects) occur about every 200,000 years, while major collisions probably occur about once every 30 million years. The collisional frequency of comets with the Earth must have been very much higher 4,500 million years ago than it is today. Indeed, the surface of the primitive Earth must have been bombarded by missiles to a degree that precluded any peaceful continued existence of life in the earliest times.

The problem for those who believe life to have arisen by spontaneous generation on the Earth is now seen to be acute. The origin appears to be forced back beyond 3,800 years ago, into an era of serious geological disturbance, which seems as if it must have been generally inimical to life. Nor can one plausibly argue that the rain of cometary missiles ceased well before 3,800 million years ago, because this is close to the oldest age which has been obtained for any area of the lunar surface. The indication is that the Moon experienced severe bombardment until about 3,800 million years ago, and if this was so for the Moon it must have been even more surely so for the Earth. The Isua rocks appear therefore to date from about the earliest time when the Earth may be said to have had a tolerably stable crust.

Another method has been used to search for traces of ancient life. It is well-known that carbon has two stable isotopes, written as ^{12}C and ^{13}C, with ^{12}C about ninety times more abundant than ^{13}C. (Isotopes of the same chemical element differ in the number of neutrons contained in the central atomic nucleus, but not in the number of protons or in the number of electrons.) Photosynthetic organisms which convert carbon dioxide from the atmosphere into biological material have a slight preference for $^{12}CO_2$ compared to $^{13}CO_2$. Thus biologically-generated hydrocarbons and carbonates have a small deficit of ^{13}C compared to hydrocarbons and carbonates generated by inorganic processes. Finding a deficit of ^{13}C in such materials within ancient sediments can therefore be taken as an indication that photosynthetic life was present when the sediments were laid down.

C. Walters, A. Shimoyama and C. Ponnamperuma used this method for investigating rocks from the Isua series. Reporting their results in the autumn of 1979 at a meeting of the American Chemical Society, they argued that it was indicative of the presence of photosynthetic activity. In a broadcast interview for the Sri Lanka Broadcasting Corporation in January 1980, Ponnamperuma was more positive:

> ... so we have now what we believe is strong evidence for life on Earth 3,800 thousand million years [ago]. This brings the theory for the Origin of Life on Earth down to a very narrow range. Allowing half a billion years (for the disturbed conditions described above) we are now thinking, in geochemical terms, of instant life. . . .

In view of the enormous difficulty of building the correct amino acid sequences of the enzymes, histones and the other complex biochemical substances discussed in chapter 2, the concept of 'instant life' hardly seems attractive. If, however, one is forced to swallow an improbability as great as we calculated in chapter 2, one might better swallow it instantly, instead of in a long drawn-out agony over an extended period of time.

In our view, life on the Earth probably had many beginnings. With living cells from cometary sources constantly showered over the Earth's surface, many would manage to survive temporarily, later to become extinct, and then, perhaps, to become re-established again. Thus we are not committed to a single line of development. Eukaryotic cells and prokaryotic cells could arrive together and begin their residence here at the same time, or they could begin in either order. Preceding the Isua rocks, there could have been a number of abortive starts, even in the highly disturbed conditions of the first half-billion years of the Earth's history. Nor in the aeons following the Isua rocks do we require either prokaryotes or eukaryotes to have maintained themselves continuously. Initial failure does not imply total extinction. Our theory provides for many second chances.

6
The evolutionary record leaks like a sieve

Figure 6.1 shows in a broad sense the life-forms that have inhabited the Earth from the Isua rocks of 3.8 billion years ago to the present day.

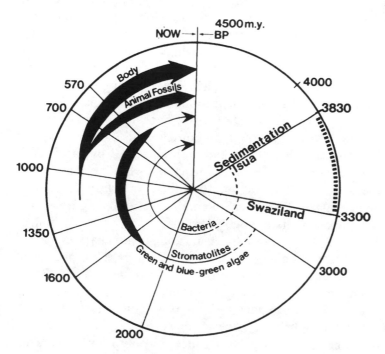

Fig. 6.1   Biological clock of Earth's fossil record

Although our purpose in this chapter is to discuss in detail the most recent 500 million years of the evolutionary record, it is interesting to begin with the more general question of why it took so long for multi-celled life to start its evolution.

The time-scale of Figure 6.1 is very long. We must therefore contemplate the possibility that changes have occurred to both the Earth and the Sun. Let us take the Sun first. Astrophysical calculations show the Sun to have been less luminous in the past than it is now, by 20 to 25 per cent at the time the oldest known terrestrial rocks were laid down. If the Earth's atmosphere had then been the same as at present, the average world temperature would have been lowered by about 40°C, a half of the lowering being a direct effect of the weakened Sun, the other half being the consequence of less water vapour evaporated from the oceans (which would largely destroy the well-known 'greenhouse effect' produced by the water vapour).

At such a markedly lowered temperature, the Earth would have been in a glacial condition with the oceans frozen. Geological evidence, however, shows the oceans were not frozen. Indeed, geological evidence suggests the oceans were a great deal warmer than they are now, with an average temperature of at least 50°C compared to the present-day 15 to 25°C. Hence we can safely conclude that the terrestrial atmosphere was not the same in the remote past as it is now. In particular, we can say that the terrestrial atmosphere 3.8 billion years ago must have produced a considerably larger greenhouse effect than it does at present.

Of the several ways that a strong greenhouse effect might have come about in the earliest times, the most plausible appears to us to be from a high atmospheric concentration of carbon dioxide, which was then gradually depleted over the succeeding Archean and Proterozoic eras. Depletion can occur in two ways, in the formation of limestone and other carbonate rocks, and through the photosynthetic action of living cells, mostly blue-green algae. The latter operate in the following manner.

In carbon dioxide there are two oxygen atoms to one of carbon, as the chemical formula CO_2 shows. In sugars and polysaccharides there is essentially only one oxygen atom to each carbon atom. Thus photosynthesis, which turns atmospheric carbon dioxide into sugars and polysaccharides, detaches one of the two oxygen atoms in the CO_2. The resulting freed oxygen atoms join up in pairs to form oxygen molecules, O_2, which then go into the atmosphere in place of the carbon dioxide.

The sugars and polysaccharides produced by photosynthesis are not deposited in a free form out in the open like the honey of bees. They constitute an integral part of the living cells that generate them. When the cells die it is usual for the reverse process then to take place. The dead biomaterial becomes oxidized, which is to say the carbon atoms pick up a second oxygen atom, and so return to their original form as carbon dioxide. If this return process is exactly a hundred per cent efficient, there is no permanent depletion of the atmospheric carbon dioxide. The carbon dioxide is simply borrowed from the atmosphere for a while, and then returned to it. Occasionally, however, dead biomaterial becomes suddenly covered by inert material. Out of contact with atmospheric oxygen, it is then no longer oxidized and there is no return of the borrowed carbon dioxide.

To the extent that dead biomaterial is slowly buried, for example by landslides or by coastal vegetation being inundated by a sudden rise of sealevel (this is how most of the coal that is currently being mined was formed), carbon dioxide is progressively removed from the atmosphere. In its place an atmospheric oxygen excess is gradually built up.

From the fact that the present-day atmosphere does indeed contain an oxygen excess this argument is seen to be correct in principle. Moreover, the atmospheric oxygen excess generated over the long ages of the Archean and Proterozoic was almost certainly far larger than we find today, because much of the oxygen has been taken up in oxidizing inorganic material, especially in oxidizing ferrous iron to ferric iron in rocks and soils.

It is a consequence of this line of reasoning that there must be a great deal of buried biomaterial inside the Earth, not just near the surface where we find economically workable reserves of coal, oil and gas, but at much greater depths. Diamonds, which form from graphite at high temperature and pressure, appear to have ascended with almost explosive rapidity (in a kind of super-volcanic blast) from depths greater than 150 kilometres, which clearly proves the presence of carbon at great depths—indeed, deep-lying carbonaceous material is probably necessary to generate volcanic gases, and perhaps even to serve as the cause of major Earth movements.

With the atmospheric carbon dioxide slowly declining, the greenhouse effect would weaken and the general world temperature would fall correspondingly. The carbon dioxide content of the atmosphere, however, did not fall completely to zero, but it fell until it could be maintained by volcanic activity. At this stage, the temperature had declined sufficiently for multi-celled life-forms to begin their evolution.

An explanation of Figure 6.1 would therefore seem to be that the early temperature of the Earth was simply too high for anything other than single-celled forms to exist. Both bacteria and the blue-green algae can survive at 70° C or more, and there was never any problem for them. Some microfungi can survive up to about 60°C, and these, as we saw in chapter 5, also existed in the early history of the Earth. But for protozoa and other animals, with temperature upper limits in the range of 45 to 50°C, the early Earth was too hot. It was necessary in their case to wait until sufficient cooling occurred. Cooling took a long time, which is why complex life-forms did not appear for so long, indeed until the sector at the upper left of Figure 6.1.

Figure 6.2 has been adapted from David Attenborough's *Life on Earth*.[1] It shows the higher taxonomic categories of plants and animals, without claiming that any connections have been discovered between them. Authors of texts in biology are often so convinced that such connections existed that they cannot refrain from drawing an evolutionary tree with all the life-forms of Figure 6.2 derived by a system of branches from a single ancestral trunk. Indeed, if one believes in Darwinism it has to be so, and in many people's minds this puts the matter beyond all doubt. It is a merit of David Attenborough's very fine book that, although he evidently believes strongly in Darwinism himself, he has avoided drawing connections which have not been established in the fossil record, although the lines have been curved in Figure 6.2 to suggest common origins.

It used to be said that the fossil record is patchy, and therefore very incomplete, and one still finds this old excuse appearing in modern biological texts. Yet by consulting geologists one learns otherwise. For example, G.M. Bennison and A.E. Wright write as follows:[2]

> Over a century ago Charles Darwin expressed disappointment that the fossil record provided less support for the theory of evolution than might be hoped for. Since that time vast numbers of fossils have been found and many new species and genera described, although the number of examples of fossil lineages which demonstrate evolutionary changes in detail is still small.
>
> How adequately represented [in the fossil record] are organisms which formerly existed on Earth? A.B. Shaw has examined the statistical probability of organisms being found in the fossil record.[3]

[1] *Life on Earth*, Collins/BBC, London, 1979, 310-11.
[2] *The Geological History of the British Isles*, Edward Arnold, London, 1969.
[3] A.B. Shaw, *Time in Stratigraphy*, McGraw-Hill, New York, 1964.

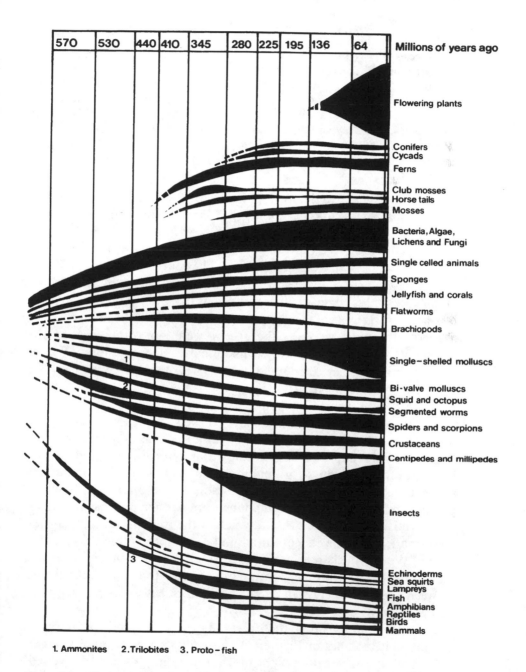

570	530	440	410	345	280	225	195	136	64	Millions of years ago

Flowering plants

Conifers
Cycads
Ferns

Club mosses
Horse tails
Mosses

Bacteria, Algae,
Lichens and Fungi

Single celled animals

Sponges

Jellyfish and corals

Flatworms

Brachiopods

Single-shelled molluscs

Bi-valve molluscs
Squid and octopus
Segmented worms

Spiders and scorpions

Crustaceans

Centipedes and millipedes

Insects

Echinoderms
Sea squirts
Lampreys
Fish
Amphibians
Reptiles
Birds
Mammals

1. Ammonites   2. Trilobites   3. Proto-fish

Fig. 6.2 Simplified forms of life—the widths of columns giving an identification of the abundance of species (From D. Attenborough—*Life on Earth,* Collins/BBC, 1979)

It is clear that, if one individual in a million of a particular species was fossilized, there is a high probability verging on certainty that if the species survived for only a million years specimens would be found. With marine dwelling animals the chance of fossilization is probably better than one in a million and in those species known as common fossils, which often occur crowded on bedding planes, considerably better [than one in a million]. Also, as Newell has shown, even some species with only chitinous skeletons or with woody tissue are not uncommonly preserved: this is especially true in impervious strata where they presumably have less chance of being subsequently destroyed.

Reference back to the work of Professor Pflug on micro-organisms in the Isua rocks of Greenland gives an insight into the delicate techniques that can now be brought to bear in deciphering the fossil record. Moreover, there are now so many workers in the field that nothing clear-cut can have been missed. The conclusion from geology is that *either* the lines of Figure 6.2 were always separate *or* lines branching off an ancestral stock made their transitions to new forms so quickly that it is not surprising geologists have been unable to find evidence of the changes. In other words, either there were no transitions or the transitions were so rapid as to be analogous to quantum jumps, where one sees atoms either in an initial upper level or a later lower level but never half-way between.

Faced with the persistent and increasing difficulty of the silence of the fossil record, evolutionists have turned from the field to the laboratory. Already in chapter 1 we discussed the polypeptide chain (with a heme group attachment) known as cytochrome *c* and we saw how, from generation to generation of a species, neutral mutations accumulate in the positions of a fraction of the amino acids in the polypeptide chain. (Neutral mutations were defined as those for which any selective advantages or disadvantages the mutations might convey were smaller than the effect of random fluctuations within the breeding group.) We saw in chapter 1 that in an evolutionary picture more neutral mutations of cytochrome *c* can be expected to have accumulated between species widely separated in the so-called ancestral tree than between nearby species.

If one could be certain that there really was an overall evolutionary picture, that the lines of Figure 6.2 are really derived from a common stock, then the expectation of the previous paragraph could be inverted. The number of differences in the amino acid chain of cytochrome *c*, determined in the laboratory, could be regarded as a measure of the separation distance of various species in the evolutionary tree. Thus the determination of such differences for many species could be used to reconstruct the whole evolutionary process. Figure 6.3 shows the tree obtained from this argument by M.O. Dayhoff, C.M. Park and P.J. McLaughlin.[4]

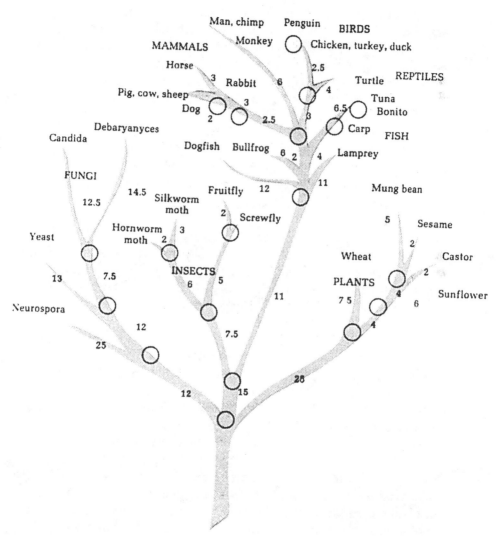

Fig. 6.3 Phylogenetic tree showing the presumed evolutionary connections for cytochrome *c*. Each circle represents the sequence of a cytochrome *c* deduced to be ancestral to all species higher in the branches leading from that circle. The figures represent the number of amino acid residue differences per 100 from the ancestor. Thus, mung bean cytochrome *c* differs from its ancestor by 5 residues per 100 (From A.L. Lehninger, *Biochemistry*, Worth Publ. Inc., 1976)

[4] *Atlas of Protein Sequence and Structure*, vol. 5, National Biomedical Research Foundation, Washington, DC, 1972.

One should not be deceived, however, by the elegance of this result into thinking Figure 6.3 proves the existence of an evolutionary tree. What is shown is that if a tree existed then it was like this.

The numbers marked along the branches of Figure 6.3 are not the actual differences of the amino acids, they are the experimental numbers multiplied first by 100 and then divided by the total number of amino acids in the cytochrome c chain, the division being rounded off to the nearest half (0.5). Examining these 'normalized' numbers, one can see that the largest values occur on just the branches where there are no present-day life-forms. Working backwards from the smaller outlying branches, where the present-day life-forms occur, it is only possible to reach the common trunk of the tree if one postulates large numbers of amino acid changes on the lower branches. This could always be done. Even if there had been no evolutionary tree giving rise to the spreading outer branches, it would always be possible to construct one by this method.

Suppose A, B, C, . . . , each to be a group of species, but with each group separate in its origin from the others. Let there be a genuine evolutionary connection, however, among the species of A, a connection among those of B, and so on. For each of A, B, C, . . . , a separate outer set of branches of the kind marked Fungi, Insects, Mammals, Birds, . . . , in Figure 6.3 could be constructed. Then whatever the jumps may be in the amino acid structures going from A to B to C, ... , lower branches leading to a hypothetical (and false in this case) ancestral trunk could always be found, simply by postulating sufficient amino acid changes to have occurred on the lower branches.

But could one have evolution within A, within B, . . . , and not have evolution connecting A to B to C, . . . ? Why not? With living cells raining constantly down upon the Earth, many distinct evolutionary lines could conceivably have come into being. Rather is it that with A ≡ Fungi, B ≡ Insects, C ≡ Mammals, . . . , we may have conceded far too much to the usual evolutionary picture. Figure 6.4 is a remarkable diagram showing the fossil record for the various orders of winged insects. Just as there are no observed connections between the major categories of Figure 6.2, so there are no connections on the much lower taxonomic level of Figure 6.4.

Insect wings occur in pairs, developed out of the triply segmented tubular structure known as the thorax. They are stiffened by veins in which there is a characteristic pattern of air channels used for respiration—i.e. along which atmospheric oxygen diffuses to the interior regions of the insects. These 'tracheae' and the veins which contain them

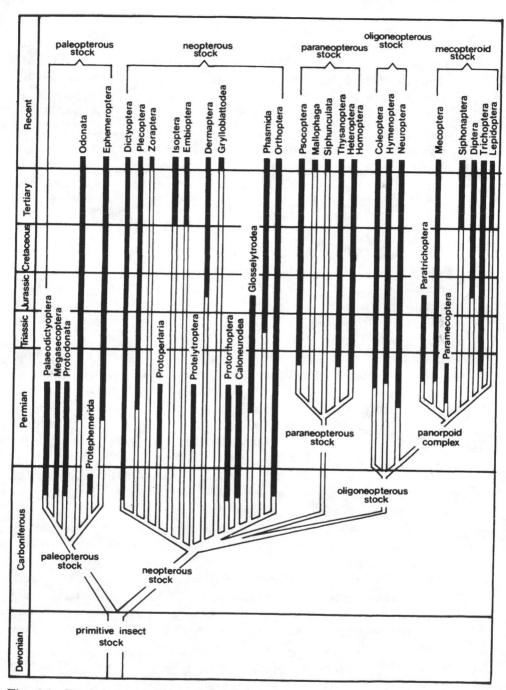

Fig. 6.4 Fossil insects have been found only along the solid segments, not along the presumed connections (From *Encyclopedia Britannica,* 1977)

have a consistency of pattern from one insect order to another which suggests to entomologists that all the lines of Figure 6.4 were derived from a common stock. However, the chitinous bodies of insects enable them to be well-preserved in the fossil record, as was remarked by Bennison and Wright in the quotation given above. (This contention is well-proven by the excellent condition of many fossil specimens within the separate orders of Figure 6.4.) The lack of specimens from the presumed lower, connecting branches of Figure 6.4 is therefore hard to understand, especially in view of the very large insect populations. It is particularly remarkable that no forms with the wings at an intermediate stage of development have been found. Where fossil insects have wings at all they are fully functional to serve the purposes of flight, and often enough in ancient fossils the wings are essentially identical to what can be found today. Nor are there intermediate forms between the two kinds of wings, those of the Paleoptera held aloft or permanently at the side as in mayflies and dragonflies respectively, and those of the Neoptera with a flexing mechanism enabling the wings to be folded back into a resting position across the abdomen.

Insects fly by flapping their wings. At each flap, the wing blade is twisted from elevation to depression, so producing the aerodynamic effect of a rotating propellor. Flight direction is controlled by changing the angle through which the wings are twisted. Butterflies have an exceptionally slow wing-beat, about 10 per second, whereas mosquitos have a rate of 500 or more per second, thereby producing the high-pitched whine that strikes abject terror into the hearts of those of us who are sensitive to their bites. It is impossible for muscles by themselves to extend and contract in only 1/500 of a second. The wing beat of mosquitos is therefore controlled not overtly by the muscles, but by the mechanical oscillation frequency of the thorax.

Winged insects fly by taking advantage of the same aerodynamic principles as man-made aircraft, and jumping insects leap by taking advantage of the same elastic principles as a man-made catapult. Muscles are used to compress an elastic pad composed of the protein resilin. This is done slowly and is analogous to the winding of an ancient Greek catapult of war. When the insect is ready, a catch is released and the insect is propelled, exactly like the missiles by which Alexander the Great made his spectacular conquests.

Entertaining as this may be, the prognosis for Darwinism is now very poor. We can only explain the absence of intermediate insect forms in the fossil record *either* by supposing the different insect orders to be of separate origin *or* by arguing that the divergencies from the common stock indicated at the base of Figure 6.4 took place with extreme rapidity. Only the second of these possibilities is consistent with Darwinism, yet rapid evolution is just what Darwinism cannot achieve. The reader will recall from chapter 1 that so-called 'point mutations' of the genes on DNA take place only very slowly, far too slowly to be of relevance here, unless one could argue that changing only one or two genes was sufficient to produce the diversity of Figure 6.4.

We would also need to argue that only a small number of genes was involved in producing the major divergences of Figure 6.2 and 6.3. This would make all of evolution contingent on only a handful of genes, which is surely absurd. Indeed, the contention is not merely absurd, it is grossly untrue. Figure 6.5 shows the varying quantities of genetic material (in number of DNA bases) of the major taxonomic categories, which are very much the same as those of Figure 6.2. Evidently the various categories differ from one to another very greatly in their genetic content. It came as a great surprise in biology that a handful of fungal cells contain as much DNA as a man, and that the salamander contains a great deal more.

Figures 6.6 to 6.10 show the scheme of vertebrate evolution as it is usually supposed to occur. These diagrams include many divergencies which have not been proved by the fossil record, so if they occurred the transitions must again have been rapid. The diagrams are therefore highly conjectural. It has been through the device of presenting such diagrams with the presumed connections drawn in firm solid lines (unlike the broken lines used in Figure 6.2) that the general scientific world has been bamboozled into believing that evolution has been proved. Nothing could be further from the truth. Consider, for example, the monotremes, an order of lower mammals comprising the duckbills and echidnas (spiny anteaters).

From Figure 6.10 it is easy to condition oneself to the belief that the monotremes are known to have existed since the Jurassic period, about 150 million years ago, when they emerged from a mammal-like reptile. Yet the *Encyclopaedia Britannica* begins its article on monotremes with the following paragraph:[5]

> Only two monotreme types, the platypus and the spiny anteaters, are known. The earliest monotreme fossils come from the Australian Pleistocene, and they are essentially the same as the living forms, which range in body length from about 30 to 80 centimetres and weigh from one to ten kilograms.

[5] *Encyclopaedia Britannica*, vol. 12, 384.

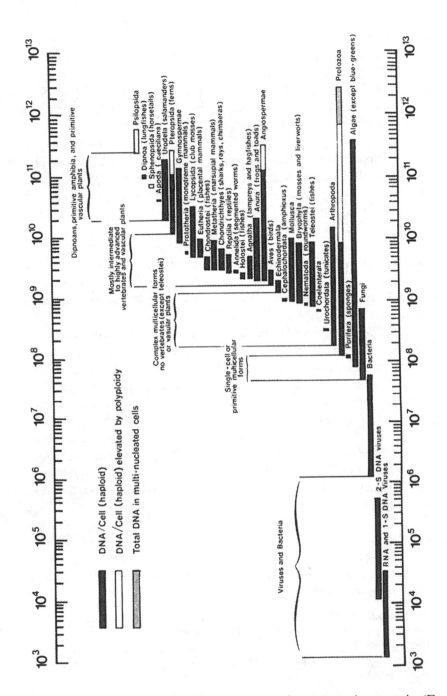

Fig. 6.5   The range of nucleotide content in major taxonomic categories (From B. John & K.R. Lewis, *Chromosome Hierarchy*, Clarendon Press, 1975)

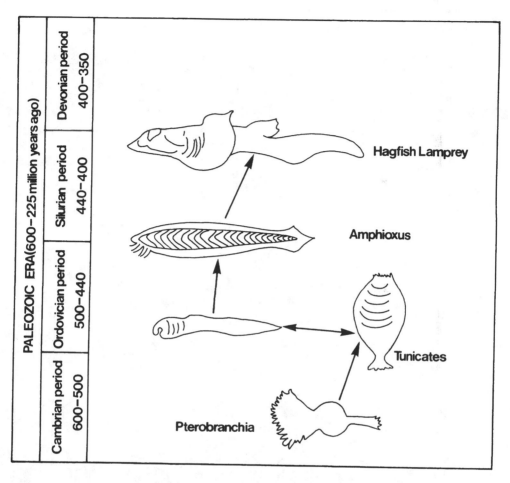

Fig. 6.6 The emergence of the first vertebrates from the larval forms of tunicate-like creatures

Since the Pleistocene is usually considered to have begun only about two-and-a-half million years ago, the situation so far as the fossil record goes is that the monotremes are exceedingly modern creatures with an antiquity going nowhere near the 150 million years back to the Jurassic. We are not concerned here to argue that the monotremes are indeed modern creatures. Our point is that the evolutionary linkages shown in the above diagrams are conjectural. The factual evidence is overwhelmingly confined to lines of creatures that do not change very much from generation to generation, as, for example, the various orders or insects in Figure 6.4. Wherever one would like evidence of major changes and linkages, such as those drawn in Figure 6.6 to 6.10, the evidence is conspicuously missing from the fossil record.

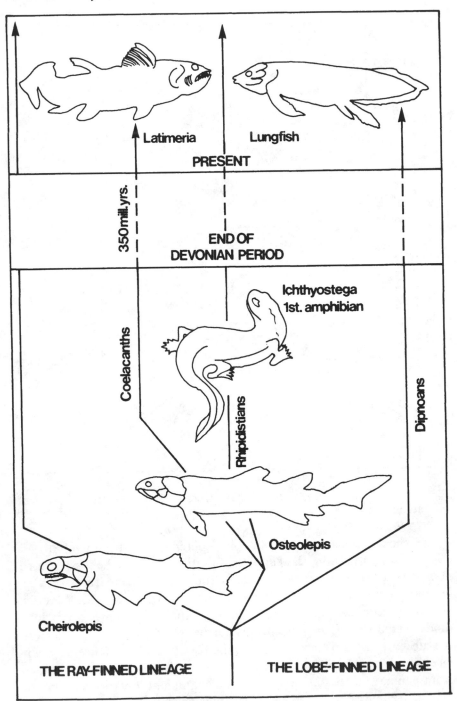

Fig. 6.7 The evolution of fish

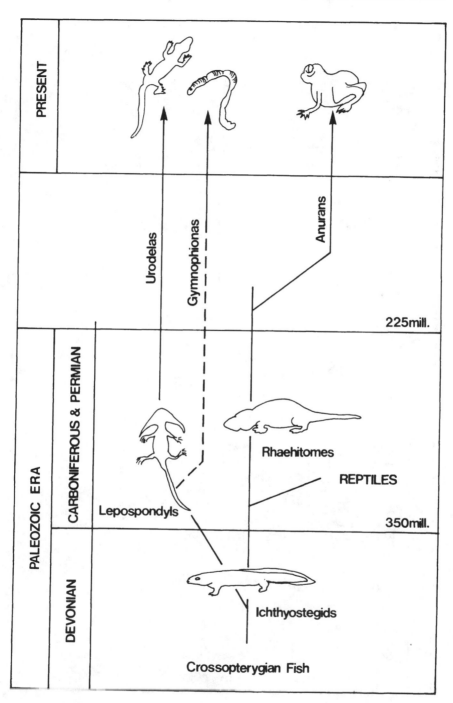

Fig. 6.8   The evolution of amphibians

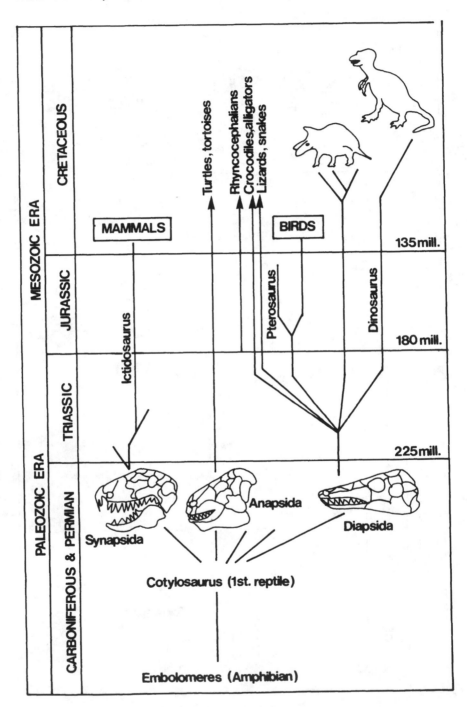

Fig. 6.9 The evolution of three sub-classes of reptiles and their descendants

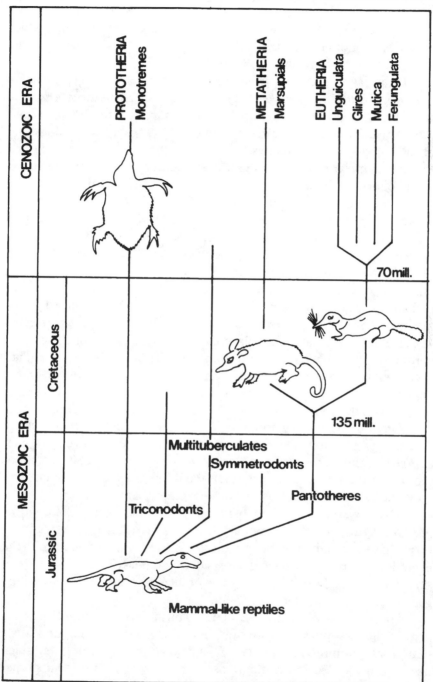

Fig. 6.10 The evolution of monotremes, marsupials and placentals from mammal-like reptiles

The problem for Darwinism is compounded by Figure 6.5, which shows that major changes and linkages are associated with large changes of the quantities of genetic material.

Let us take a rather more detailed look at the assumed divergence in Figure 6.9 from *Cotlyosaurus,* the first reptile, into *Synapsida, Anapsida* and *Diapsida.* Because our own line of mammals is supposed to have been derived from *Synapsida,* reptilian survivors from *Synapsida* would be of more subjective interest to us than those from *Anapsida* and *Diapsida.* It is typical of the unsatisfactory nature of the usual evolutionary story that there are no reptilian survivors from *Synapsida,* although there are plenty from *Anapsida* and *Diapsida.*

So it is to the survivors of *Anapsida* and *Diapsida* that we must look for genetic comparisons with the mammals. Modern turtles, believed to be derived from *Anapsida,* have genomes that are generally about 80 per cent of mammals. However, the DNA of turtles is distributed in microchromosomes as well as in chromosomes of more or less mammalian size. Mammals do not have microchromosomes. Table 6.1 shows the situation for some survivors of *Diapsida.* The genome sizes are given as percentages of that of the placental mammals, 2n being the diploid chromosome number (the number in the cells of the body). Reptilian survivors of *Diapsida* have a genome size that averages about two-thirds of those of the placental mammals, while avian survivors have about a half, and both kinds of Diapsidan survivors have microchromosomes, which are absent in the placental mammals. The genetic differences are evidently gross, and while the gap can be closed somewhat in exceptional cases (the caiman, for example, has a genome that is some 80 per cent of the size of the mammalian genome), the differences are not small.

The two major conclusions of this chapter are:

(1) The absence from the fossil record of the intermediate forms required by the usual evolutionary theory shows that *if* terrestrial life-forms have evolved from a common stock, the major branchings in the evolutionary tree must have developed very quickly.

(2) The major branchings, *if* they occurred, were accompanied by genetic changes that were not small.

These conclusions dispose of Darwinism, which cannot produce genetic changes quickly, as we saw already in chapter 1. The situation is that, if we retain the evolutionary scheme favoured by biologists, which has been illustrated in the diagrams of this chapter, some means other than the usual Darwinian picture must be found to drive the evolution. The alternative is to suppose that the major categories of Figure 6.2 and

Diapsidan reptiles

(From S. Ohno, *Evolution by Gene Duplication,* Springer Verlag, New York, 1970)

Order	Species	2n number	Genome size %
Squamata	Chameleon lizard *Anolis carolinensis*	36 (24 micro)	60
	Alligator lizard *Gerrhonotus multicarinatus*	46 or 48 (26 micro)	65
	Boa constrictor *Boa constrictor amarali*	36 (20 micro)	60
	South American Xenodon *Xenodon merremii*	30 (14 micro)	67
	South American jararaca *Bothrops jararaca*	36 (20 micro)	60

Birds

Order	Species	2n number	Genome size %
Galliformes	Chicken *Gallus gallus domesticus*	78 (about 60 micro)	45
Columbiformes	Pigeon *Columba livia domestica*	about 80 (about 60 micro)	55
Psittaciformes	Parakeet *Melopsittacus undulatus*	about 58 (36 micro)	44
Passeriformes	Canary *Serinus canarius*	about 80 (about 60 micro)	59

Table 6.1 Genome size as percentage of placental mammals

even the insect orders of Figure 6.4 have always been distinct, and that most of the cross-linkages of Figures 6.6 to 6.10 are illusory. In the next chapter, we shall arrive at a decision between these two possibilities.

The second possibility is curiously reminiscent of the special creation theory advocated in the second half of the eighteenth century by William Paley. After emphasizing that plants and animals are remarkably well adapted to the environments in which they live, Paley likened the precision of the living world to a beautifully made watch. He then argued that, just as a watch owes its origin to a watchmaker, the world of Nature must owe its origin to a Creator, *God.* His work *A View of the Evidence of Christianity* was highly influencial. Given initially as a series of lectures in the University of Cambridge, the book became required reading in that University. Indeed, up to the twentieth century an examination, popularly known as 'Paley's Evidences', had to be successfully passed by all graduates of the University.

In 1835 and 1837 Edward Blyth published two papers in which he considered the effects of natural selection. He argued that once species were adapted to their environment, natural selection would prevent them becoming disadapted. He also argued closely along the lines of the quotation from Darwin given in chapter 4: 'On the average every species must have some number killed year with year by hawk . . . .' Blyth saw this argument clearly, long before Darwin wrote it in his notebook, but he felt that it could only improve the adaptation of an already adapted plant or animal. Getting the adaption there in the first place remained a problem, and for this Blyth found it necessary to take a position not much different from Paley's. The problem of getting the 2,000 necessary enzymes there in the first place remains to this day, as we discussed already in chapter 2.

What Darwin, and also Alfred Russel Wallace, did nearly a quarter of a century after Blyth was to assert that natural selection would indeed get the adaptation there in the first place, a position which Blyth had considered and rejected. The assertion was without proof, although the scientific world has been persuaded into thinking that exhaustive proofs were given in *The Origin of Species* (1859). What we are actually given in Darwin's book are very many changes of adaptation by already adapted species, of which there had never been any real cause for argument since Blyth's papers in 1835 and 1837. The key issue, namely that origins from scratch cannot be explained in the same way, is not dealt with at all.

The speculations of *The Origin of Species* turned out to be wrong, as we have seen in this chapter. It is ironic that the scientific facts throw

Darwin out, but leave William Paley, a figure of fun to the scientific world for more than a century, still in the tournament with a chance of being the ultimate winner. We shall see how the argument goes in the remaining chapters.

7
Cosmic genes

We saw in the preceding chapter that *if* the usual scheme of biological evolution, as illustrated in Figure 6.6 to 6.10, is to be maintained, then the biggest evolutionary transitions must take place swiftly. We likened the situation to a quantum transition, with atoms seen by an observer both before and after the transition but not seen in mid-transition. The analogy cán be taken a little further. The average time for an atomic transition (reciprocally related to the so-called transition probability) is considerably longer than the oscillation period of the emitted radiation. In a similar way, what we have been describing as a 'swift' evolutionary transition may yet be much longer than the generation of the plant or animal in question. By 'swift' we simply mean too fleeting to be imprinted in the fossil record. This would still permit thousands of generations to elapse, but not the millions of generations that would be needed to establish significant evolutionary changes through copying-error mutations altering the disposition of amino acids along polypeptide chains.

We also saw in the preceding chapter that major evolutionary changes would need to be accompanied by considerable jumps in the gross amounts of genetic material contained within the cells of the changing categories. We have now to ask whether such jumps could occur in no more than a few thousand generations through the operation of 'orthodox' biological processes. There are two possibilities, polyploidy and tandem gene duplication, which we shall begin this chapter by considering.

Polyploidy consists of a duplication of the entire genome. The offspring gets a double set of chromosomes from the parent(s). For species capable of self-fertilization—some species of insects, fish and lizard, but more particularly plants—it is easy to see how a polyploidal individual can propagate its abnormality by budding or 'cloning', to use the fashionable word. But for animals with distinct sexes a polyploidal individual without a polyploidal mate is doomed to sterility. Even if one postulates that a doubling of the genome happens contemporaneously for several individuals of both sexes, polyploidy fails for mammals, birds and reptiles, because it has the effect of destroying the sex-determination mechanism of these categories. Thus polyploidy can be of no help towards explaining Figure 6.9 and 6.10.

Occasionally the enzymes that copy DNA may copy twice over a length of DNA containing a particular gene. If this happens during the production of a sperm or ovum, an individual with two copies of the gene side by side, in tandem, may come to be born, the two copies being contained not only in the sex cells of the new individual but in all its body cells. If, moreover, it happens that the gene in question produces a protein which the species could well benefit from having more of, the new individual with a double dose of the protein will have an advantage over other single-gene individuals and so the double gene will tend to spread itself by natural selection. Even if the double gene conveys no particular advantage it may still spread by random effects, provided it conveys no disadvantage either, especially if the breeding group to which the individual belongs is small.

With the double gene established within a breeding group, the way is open to the generation of individuals possessing more and more adjacent copies of the gene. This happens through occasional 'unequal cross-overs' in the production of sperms and ova. Thus for an individual that inherits the double gene from both parents, unequal cross-over produces sex cells with either three copies or only a single copy of the gene. If the three copies again confer an advantage, it too will tend to spread by natural selection, and unequal cross-over can then produce up to five copies. For an individual inheriting five copies from each parent, unequal cross-over can produce up to nine copies, and so on with the gene number pretty soon increasing explosively, as it has actually done in salamanders and newts.

It is possible that tandem duplication of one or several genes could produce a marked increase in the amount of genetic material over only a few thousand generations, but it is doubtful that any marked functional diversity could arise in this way. Indeed, quite the reverse. In writing about the lungfish, S. Ohno remarks:

> By establishing such a system [tandem duplication] the organism effectively forfeited an opportunity for further evolution. In a manner of speaking, the genome became frozen, while containing enormous genetic redundancy. It is clear that in doing so such a lineage reached an evolutional dead end. It will be shown that what happened to the lungfish also happened to salamanders and newts. . . . Indeed, this side branch stopped dead at the amphibian stage.[1]

Sometimes, however, tandem duplication plays an essential role. There is need for more t-RNA molecules of the various kinds (chapter 2) than single genes can provide. But the duplication of each t-RNA gene is kept down to a reasonable number, about half-a-dozen copies in yeast cells, for example.

The genes so far considered have been 'expressed' genes, those which gave rise to the polypeptides that determine the physical properties (the so-called phenotype) of a plant or animal. We saw in chapter 1, however, that most of the DNA goes unexpressed. Among the unexpressed DNA, sequences of bases are sometimes repeated an enormous number of times, as many as a million times for certain short sequences. Figure 7.1 shows how the genome of the mouse is divided between single genes (about seventy per cent in quantity of DNA), genes with from one thousand to one hundred thousand copies (about twenty per cent), and genes with a million or more copies (about ten per cent).[2]

The function (if any) of these enormously repeated sequences of unexpressed DNA is still unclear. There is something of a mystery as to why tandem duplication has not got entirely out of hand to swallow more or less the whole genome. The fact that after millions of generations the mouse has still most of its genome devoted to single genes shows that there must be a process which intervenes to prevent such a genetic disaster from taking place.

At all events, tandem duplication does not solve the evolutionary dilemma. It might give a rapid increase in the quantity of genetic material, but it only does so by being highly repetitive, and this will not give a sequence of 'quantum jumps' in the forms of plants and animals, such as is needed to provide for the divergent evolutionary branches shown in Figures 6.6 to 6.10. Repetitions will give some changes, of course, by altering the quantities of certain proteins, but, as Ohno remarks in the above quotation, the changes are much more likely to be stultifying than to lead to adventurous new possibilities.

[1] *Evolution by Gene Duplication*, Springer Verlag, New York, 1970.
[2] From R.J. Britten and D.E. Kohne, *Science*, vol. 161, 1968, 529.

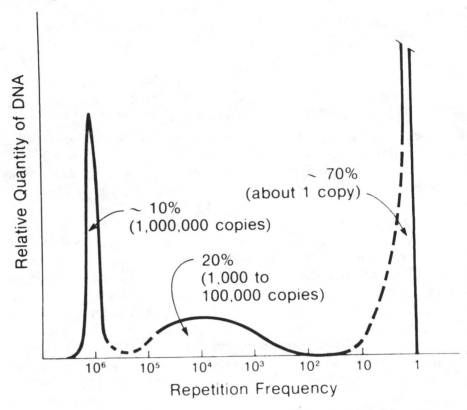

Fig. 7.1 Proportion of the total DNA with a given level of sequence repetition in the mouse. The scale of the abscissa is logarithmic (From T. Dobzhansky, F.J. Ayala, G.L. Stebbins and J.W. Valentine, *Evolution*, W.H. Freeman & Co., 1977)

Dr Ohno's book *Evolution by Gene Duplication* argues the present issues (on the basis of terrestrial theory) as fairly and as completely as any treatment we have read. He argues that polyploidy occurred in the mammalian-reptilian ancestral stock at some stage before Figure 6.9, before polyploidy became forbidden by the sex determination mechanism. The genes on only one of the duplicate sets of chromosomes were sufficient to maintain the ancestral stock, most of the genes on the other set being free to drift through the accumulation of neutral mutations.

To speed the fixing of neutral mutations, Dr Ohno requires the ancestral breeding group to have been rather small. The free genes then go in biochemical directions that have little to do with natural selection. In this way a gene can drift, even within a reptile, to a form that will be of later use to man. Indeed, the genes are supposed to have drifted to a configuration which determined the later evolution of Figure 6.10, and the still later evolution of Figure 7.2, not by the need to adapt to the immediate environment, but by chance. The chance anticipation of later need continued even up the last stage of Figure 7.2. Thus in his concluding pages Dr Ohno remarks:

> Did the genome of our cave-dwelling predecessor contain a set or sets of genes which enable modern man to compose music of infinite complexity and write novels with profound meaning? One is compelled to give an affirmative answer . . . . It looks as though the early Homo was already provided with the intellectual potential which was in great excess of what was needed to cope with the environment of his time.

Dr Ohno was thus led by his resolute respect for the biological facts to what seems to us a non-Darwinian position almost as marked as our own. The facts, interpreted within a terrestrial theory of the origin and evolution of life, force one to suppose not only that chance faced up to the incredibly minute probability of discovering the enzymes and other basic biochemical substances, a probability we calculated in chapter 2 to be less than $10^{-40,000}$, but that chance mutations also produced genes which were to prove capable of writing the symphonies of Beethoven and the plays of Shakespeare. This is the position to which one is inevitably led by following an Earth-bound theory, a position that we believe to constitute a *reductio ad absurdum* disproof of that theory.

The concept of 'gene' has so far been restricted in this book to what are sometimes called 'structural' or 'functional' genes. These are genes that give rise, when expressed, to polypeptides that do something or that serve some clearly defined physical purpose. Cytochrome c has the clear physical purpose of acting as an electron donor, so that the genes which carry the information for building cytochrome c may be said to exist for a functional purpose. So too are the genes which give rise to hemoglobin and to the two thousand enzymes. The t-RNA molecules serve actually to build polypeptides, which then go on to do something. The genes giving rise to the t-RNA molecules may therefore be said to be structural.

This, however, cannot be all. The battery of structural and functional genes cannot simply be left to themselves, otherwise the situation would be like a number of inanimate cars in a parking lot, each one of them functionally and structurally complete, but with the system static. Or if the inanimate cars were all started up, put into gear, and then let loose, total chaos would result.

It is interesting to watch the activity in a large crowded parking lot at the end of a big sporting event. At first sight, especially in the early

Cosmic genes

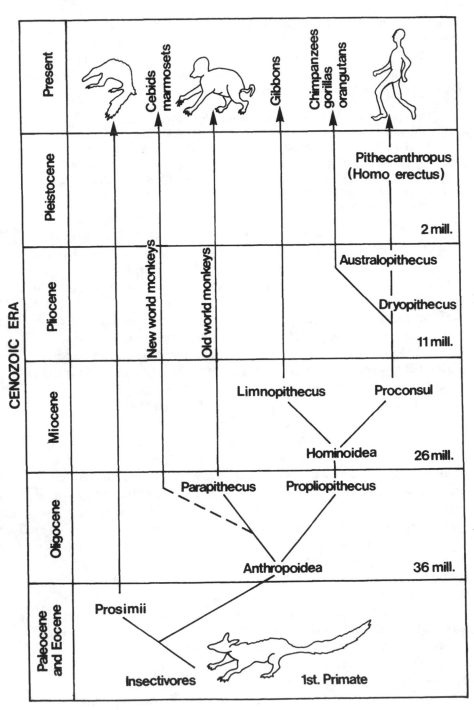

Fig. 7.2 The family tree of primates

moments as almost all the drivers are simultaneously attempting to get away, the situation does seem like total chaos. Yet as the minutes tick by the coordinated edging of the drivers somehow manages to produce a very gradual thawing of the situation, with the result that in about an hour the place will be emptied of its cars. Indeed, instead of the cars being jammed together into a woeful tangle fitted only for the wrecker's yard, they almost all end up safe and sound in their owners' garages. It has been, of course, the coordinated programs in the minds of the drivers which made this apparently miraculous outcome possible.

In a similar way there must be a program that directs the activity of a living cell. The question is what decides this program, and where inside the cell are the instructions for it located? To take the easier second part of this question first, while biologists are generally agreed that such instructions must exist, the situation concerning their location is indefinite. The usual disposition is to suppose that the location is on the chromosomes. If so, a possible location would be in the so-called nucleolus, a chromosomal region that appears decisively during the process of cell division and which would seem to preserve its identity in that process.

Cell division is a violent electro-mechanical affair, and the location of the cell program must be such as to preserve its integrity through the vicissitudes of both mitotis (normal cell division) and meiosis (a more complex double division leading to the production of sex cells).

The logic of positioning the cell program in the chromosomal DNA would be somewhat defective, however, for DNA alone is inert, and this would seem to be just the one thing which the cell program should not be. If the cell program were on the DNA, something else would be needed to interpret it, and then this 'something else' would essentially by definition become the cell program. Indeed, it is just because the DNA is so inert that it is so useful for storing already available information. Conceivably the DNA could store a copy-for-reference of the cell program, but it is hard to see the operating program being itself on the DNA. Be this as it may, the situation even for the location of the cell program is unclear, while the manner of its overall control of cell activity is quite unknown. Let us therefore turn to an analogy where the manner of operation of control programs is well-understood, to computers.

Analogies tend to have a bad name, because they are often ill-chosen, sometimes deliberately so in order to support a false argument. Yet there are situations well-known in mathematics where analogies can be exact, where the basic logic of two distinct situations is precisely the same. The physical structure of crystals and the behaviour of permutations (used by football-pool addicts) both obey the rules of the mathematical theory of groups. So do many aspects of the laws of physics. When several distinct situations thus follow the same inner abstract logic they are said to be 'representations' of that logic. We suspect that the operation of computers and the regulation of living cells are both representations of the same (or very similar) form(s) of logic.

Suppose a person who had never seen a computer, but who yet had an expert knowledge of the principles of electronics, were shown a very large computer in full operation. What would he or she make of it, given the opportunity to press switches, but not given any circuit diagrams or instruction manual?

Our imaginary investigator, if skilful enough, would discover that the computer was a vast aggregation of comparatively simple circuits, each of which performed a straightforward function, as the enzymes do in biology. By using test equipment from the black bag which investigators always carry, he or she would discover the concept of a 'bit', an electronic representation of 'yes' or 'no', and our investigator would find that the information content of the computer consisted of many rows of bits, 'words' as they are called. Pursuing the matter further, he or she would find physical locations in the computer assembly (disks or tapes) where many thousands, perhaps hundreds of thousands of 'words' were stored for reference.

All this is strikingly similar to the situation in the living cell. For disks or tapes substitute DNA, for 'words' substitute genes, and for 'bits' substitute the bases adenine, thymine, guanine and cytosine. True, there is a difference in that a computer bit has only two alternatives (yes or no) whereas a biological base has four alternatives. But by combining bits two-at-a-time even this difference could be removed (in modern practice, bits are often combined three-at-a-time to give eight alternatives—such groups of three, analogous to base triplets in biology, are called 'bytes').

The big problem for our investigator would of course be to discover what the computer was doing. Our person would see lights flickering on the console, disks spinning, tapes rotating, and every now and then yards of paper covered with words and numbers would spew out from a printer.

There is a facility in all computers whereby one can stop them for a while and then restart without losing anything of their operation. While stopped, it is possible to hand-step a computer program, causing the program instructions to be executed one by one, a procedure that is emphatically not recommended in instruction manuals. Our investigator is at so much of a disadvantage, however, that we will allow this facility, which then would permit the computer to be operated in extreme slow motion.

So the investigator might discover that the computer adds, subtracts, multiplies and divides numbers which have been represented by 'words'. He or she could discover that the computer copies 'words' from one part of itself to another, and it would be found (given much hand-stepping!) that the computer has many repetitive sets of operations, as when it determines the logarithm of a number or the tangent of an angle. Thus the investigator would be led to the concept of a subroutine, a separate self-contained part of the program, closely analogous to the expression of a gene in biology.

Yet there would be nothing in all this to answer the primary question of what the computer was doing, and likely enough no amount of probing and prying would succeed in answering this question. If, however, we were now to invite the investigator to stand back in the shadows and watch while the computer was operated by its normal staff, puzzlement would soon abate. The investigator would come to realize that the computer had no unique program, that it had no single function, but would operate on any program, provided only that the program was suitably arranged to match its own inherent mode of operation. The investigator would see many programs coming from different outside users of the computer being executed, and he or she would at last realize that of course it was the external users who were supplying the programs.

Following this little fantasy, let us go on to consider a larger fantasy. Suppose you were the director of a cosmic life agency with the aim, not only of spreading life wherever there was an environment that could support it, but of spreading life that could develop the greatest measure of complexity consistent with the physical properties of each individual environment. How would you go about your job?

One method would be to make a grand tour of the universe; carrying with you a sample kit of all possible life-forms and depositing as you went the forms most suitable for each individual environment. You are required to obey the laws of physics, however, and a little calculation soon shows that such a procedure, using even the most efficient transport, would take an impossibly long time. Besides which, except for extreme addicts of space travel, this method would become impossibly boring even before one was a tenth part of the way through the program, since it would be necessary to visit a total of at least 10^{20} possible environments.

Another shortcoming of such a crude system, after dumping a

particular life-form in a particular environment, is that it would be necessary to hurry along immediately to the next port of call. There would be no opportunity to nurse the fledgling creatures along, no opportunity to redirect the course of evolution in each environment. Each case would be a one-shot affair, and this would surely prove highly inefficient. Repeated visitations to each environment would again be ruled out by the laws of physics.

A much more subtle system is needed, a system that once designed will do all the work automatically for itself, leaving you free to relax, drinking mint julips in the shade of the B0 Ib star where the ever-peripatetic NASA has built your space agency for you.

The method described in chapter 4 gives all that is needed for spreading life on a cosmic scale *to start with*. But what about its subsequent development to complex forms, the phenomenon we call 'evolution'? In making your plans you might have decided to leave the evolution in each environment to Darwinism, but if so the results would be poor. Nor does this seem to have happened in the case of the Earth, as we saw in the preceding chapter. So how might you go about improving the chances for the development of complex life-forms, aimed eventually towards the emergence of intelligence?

Certainly not by attempting to guess the nature of each environment beforehand and by designing a cell with a complex program fitted to your guess. Unlike *God,* you do not see every sparrow that falls. So you build cells with simple robust programs, like yeast cells, that can manage to survive and multiply in the broadest possible range of environmental conditions. Assuming success in this first step for some particular environment, you have no wish for the cells simply to go on and on repeating themselves indefinitely. So you make the cells with the potential to switch over (once they have become established) to a variety of more complex programs.

To operate complex programs successfully, it is necessary that the cells be equipped with far more storage, far more sub-routines, than they need in the first place. In biological language, the cells must be equipped with batteries of genes that have no place in the initial functioning of the cell, but which may come to be 'expressed' later when the cell, either alone or in association with other cells, has 'evolved' to a more complex pattern of behaviour.

This gives a first glimpse of the reason why, in Dr Ohno's words: 'It looks as though the early *Homo* was already provided with the intellectual potential which was in great excess of what was needed to cope with the environment of his time.' It also gives an insight into the curious fact with which we began this book, that peas and beans contain genes for the production of blood.

You could by no means leave the program changes needed to make more complex life-forms to mere chance. Every competent space-mathematician would assure you that such a Darwinian idea had no chance of working, nothing at all would happen. Every computer expert will in fact assure you that throwing random mistakes into a computer program is no way to improve it, the probability of doing so being absurdly small, like the 10^{-40000} we arrived at in chapter 2.

Something overt must therefore be done. Now that the cells have become rooted in their new environment and are replicating themselves, so that a numerous population has been built up like the population of yeast cells discovered by Professor Pflug in the Isua rocks of W. Greenland, you can afford to take risks. You can risk destroying a number of them in order to get an improved program established. This you do by letting genetic fragments rain gently from space into the environment, each fragment being a small program in itself, intended to change a fraction of the cells by directing them as follows:

> Stop what you are doing now, add these further instructions to your program, and then continue at such and such a point.

Although the idea is simple enough in principle, it doesn't take much calculation to show that there are problems. The first trouble is that you cannot know in advance what constitutes an 'improved' program. However shrewd a guess you might make, in the last analysis 'improvement' must rest logically with natural selection in the environment in question. The problem comes from the fact that you have 10^{20} or more environments to deal with. To take account of them all you will be faced by the need for many different fragments, suggesting many kinds of change in the cell programs. What looked like a unique highroad turns out to be a maze of pathways.

In a particular environment most of the suggested changes are likely to be deleterious rather than improvements. Deleterious effects may kill cells, but since the rain of genetic fragments into the environment is only light, not more than a fraction of the cells are likely to be thus affected, and by replication the remainder soon make good the losses. It is therefore not the commonplace deleterious effects you have to worry about, but the rare advantageous cases. When improvements come, you have to seize on them avidly. If an improvement occurs only to an individual cell, it is unlikely to spread through the whole population unless the effect is gross, and the chances are the improvement will be lost. This difficulty can be avoided, however, if you design your genetic fragments to be infectious, so that they tend to spread from cell to cell in the manner of viruses or the plasmid components of bacteria.

There is a strong reason for pushing ahead as fast as possible with the improvements, even at the risk of spreading deleterious infections by a virus type of particle. You cannot afford to let too many generations pass by, otherwise copying errors will garble your precious store of cosmic genes, the ones that have not yet been expressed. When this happens, as it assuredly will in about ten million generations, all possibility of evolution derived from the initial store of unexpressed genes will be gone. In our view garbled genes are garbled genes. They do not produce the symphonies of Beethoven.

It would be a huge step to pass in ten million generations or less from a yeast-like cell to the symphonies of Beethoven, and quite certainly this did not happen on the Earth—terrestrial evolution has been of the order of a thousand times slower than this. The problem of stretching out evolution beyond ten million generations, however, would quickly be solved by your cosmic space agency experts. As well as sending in a rain of suggested program changes, send in a continuing supply of new genes.

This could be done in two quite different ways, either by supplying new genes steadily a few at a time or by providing occasional large blocks of them. Evolution in the first case would be smooth, with the new genes replacing those that became garbled, a new set arriving every few million generations. At the end of a thousand million generations the genome would thus have grown to at least a hundred times its original size, and likely enough it would have accumulated a great deal of wasted DNA. Because of the smoothness of the process, the evolutionary diversity would not be very large.

Adding sudden large blocks of new genetic information would be far more dramatic in its consequences. It would be equivalent to the upgrading of a computer configuration. To be successful, however, upgrading needs to be done with care. In computer language, the new components must be fully compatible with the old, and in the biological case this would not usually happen. Usually there would be a serious mismatch between the new and the old. But it is a remarkable property of biology, because it deals in large numbers of individuals, and because such changes as we are now considering happen only to a fraction of the total population, perhaps only a small fraction, that mismatches do not matter, except of course to the unfortunate individuals who happen to be afflicted by them. The mismatches appear as diseased cases and simply die off, leaving their places to be filled by the normal healthy majority.

What matter in biology are the exceptional cases, the cases where there is no serious mismatch. With the addition of substantial blocks of new genes, such cases are in a position to take off like upgraded computers, able to cope with problems that were quite beyond the capacity of the originals. The normal healthy majority, so necessary before as a replacement stock, has now to compete with individuals suddenly endowed with apparently miraculous new properties, and in this competition it is the previously healthy majority which then goes to the wall. The new individuals appear god-like, and it is in the very nature of being god-like that only a few are needed to replace the many.

Adding replacement genes a few at a time would permit small program improvements, but it would not permit much deviation from the previous program, which must therefore be maintained without appreciable diversity. On the other hand, the addition of large blocks of genes would create the opportunity for dramatic variability. The new programs could go in all manner of directions according to the many ways in which the large store of new genetic information could be interpreted. The effect in short would be what biologists call a 'radiance', a sudden branching of an ancestral stock into many diverse lines. To mix expressions from physics and biology, the effect of a large block of matched genetic information would be a quantum jump followed by a radiance.

These considerations clear up the big difficulty we encountered in chapter 6. We saw there that intermediate forms are missing from the fossil record. Now we see why, essentially because there were no intermediate forms. When a computer is upgraded there are no intermediate forms. The new units are wheeled in beside the old computer, the electrical connections are made, the electric power is switched on, and the thing is done. Perhaps not quite like that—there may be teething troubles. Similarly, for a few generations our new godlike individuals may not be so godlike. They may even have a rather diseased look about them, just as an upgraded computer with teething troubles seems for a while like a diseased instrument. Eventually, however, the troubles disappear, in a few days or weeks for a computer, and in a few tens or hundreds of generations in the biological case. At all events, the teething troubles are over in too short a time for them to be imprinted in the fossil record.

The story of the insects is in remarkable conformity with this picture. Thus the fossil record has many gaps. In the Devonian period, about 375 million years ago, only the collembolans, a primitive wingless form, have been found as fossils. By about 300 million years ago, however, ten insect orders are known as fossils, *but without any connections with the earlier period having been discovered*. Something critically important happened for insects about three hundred to three hundred and twenty-five million years ago, and it happened so quickly that no snapshot of what took place is found in the fossil record. Thus the story that we considered in chapter 6 for the orders of winged insects applies to the whole division of the insects.

Unless a new block of cosmic genes is made use of rather quickly in geological terms it will become garbled and useless. This tells us that evolutionary potential lasts for a while after such an acquisition, and then goes dead. Figure 7.2 shows evidence of a radiance in our own lineage about thirty-five million years ago, with a subsidiary branching occurring about seven million years ago. We saw in chapter 1 that one out of possibly twenty neutral changes of the amino acids on the α-globin chain has occurred between the gorilla and man, which measures the degree of garbling for an interval of seven million years. The same rate of about one in twenty occurring in hitherto unexpressed genes represents a moderate degree of degradation, but not sufficient to destroy future evolutionary potential. This is on the assumption that our lineage obtained pristine cosmic genes seven million years ago. Unused genes from a block of genetic material acquired thirty-five million years ago, however, would be well on their way by now to becoming dead.

Of course it is generations-elapsed rather than time-elapsed which matters in this respect. Species with short generations—insects, annual flowers, mice, hamsters—can have no remaining evolutionary potential in the present sense. Their chances for the future must rest on acquiring a further supply of genes, or on Darwinism, which is not likely to do much for them. Very long-lived species like the California redwood are the opposite side of the coin. Although the conifers are about three hundred million years old, a long generation time of, say, three hundred years gives only a million generations, insufficient to do much garbling. Thus the genome of the California redwood must be largely pristine. These splendid trees should therefore have great evolutionary potential, although in its nature the computer program for the redwood is evidently of a highly specialized form that is not easily perturbed into other channels. Similar remarks apply to the turtle, which among the reptiles is likely to have the most interesting genome. Future genetic engineers should get good mileage from both the redwood and the turtle.

The division Angiospermae, the flowering plants, emerged only about a hundred and twenty-five million years ago, and among the angiosperms there is a particularly interesting order, Scrophulariales, which is not represented in the fossil record until about sixty-five million years ago. This order contains many familiar plants—tobacco, the tomato, the potato, eggplant, the vine, foxglove, veronica, and herbaceous annuals, biennials and perennials. It comes as something of a surprise to find that it also contains some of the world's most brilliant and beautiful trees, for example the jacaranda and the white-flowered Indian cork tree. This is an example of a remarkable radiance which has not yet been greatly diminished by extinctions, even though many of its members are particularly susceptible to disease.

While there is nothing to stop a static species without evolutionary potential from being attacked by disease, by microfungi for example, there does appear to be a general correlation between evolutionary potential and susceptibility to viral diseases. Diseases of the potato and of tobacco are of course well-known. One does not often see short-lived plants and animals swept by such diseases, although again transfers of disease from one species to another do occur. It is interesting that among birds those in which influenza virus has been found—chicken, duck, turkey, quail, shearwater—tend to be long-lived species. We remarked above that program changes which are not improvements show up as disease, from which it follows that species susceptible to program changes—i.e. those still with evolutionary potential—must exhibit a greater tendency to disease than static species without evolutionary potential.

Returning to your space-agency role, there is no way in which you can know the precise state of play in each of the 10^{20} environments in your domain of influence, and so there is no way in which you can avoid genetic fragments relating to early evolutionary stages being disbursed in environments where the life-forms have long since developed past those early stages. So there is no way of preventing computers that are executing complex programs from being instructed to change to simpler programs. It follows that every species must in a few of its members be constantly receiving instructions for 'degradation' instead of 'advancement', but since a degraded plant or animal will not usually be able to compete in its environment, such backward steps will in the main disappear quickly, and unless the situation is observed in fine detail they will not usually be noticed. There is, however, one sinister case where this effect is very much noticed.

Instead of thinking of degradation instuctions being received by a few individuals in a large population of some species, think of degradation instructions being received by a few of the cells in the body of a many-celled plant or animal. Think of instructions to return, not to

a degraded plant or animal, but to the simple broad program of
individual cells, a budding program like that of yeast cells, from which
the whole thing started. A cell so instructed will soon generate a huge
family of similar primitively operating cells, because of course it finds
itself in a highly nutrient environment, namely the body of the
unfortunate plant or animal. The colony of primitive cells will grow like
the grains of wheat on the chessboard described in chapter 4, until they
become a canker or a cancer.

The phenomenon of cancer is an inevitable consequence of the
present ideas. Although the phenomenon itself is crude, the cause is
subtle. We suspect that cancer is a problem which science will not solve
until its relation to the origin and evolution of life is better appreciated.

To end this chapter we wish to return to the argument at the end of
chapter 6, to evolution from a common stock versus special creation,
where by 'creation' we now mean incidence from space. Although we
have written throughout this chapter as if the evolutionary possibility
were correct, there is no reason why some of the categories shown in
Figure 6.2 should not always have been separate. Even if there really was
a trunk to the ancestral tree, as depicted in Figure 6.3, the lower branches
of the tree could well have forked off the trunk in the cosmic setting of
Figure 4.1, long before the Earth itself was formed. Indeed, we consider it
essentially certain that no evolutionary connection has ever existed in a
terrestrial setting between so-called prokaryotic and eukaryotic cells,
between bacteria and yeast cells for example. It is even doubtful that a
terrestrial connection has ever existed between aerobic and anaerobic
bacteria, and all the main lines of chemautotrophic bacteria (those living
on quite distinct chemical reactions) are very likely also of distinct
extraterrestrial origin.

Returning to the categories of Figure 6.2 we doubt that a terrestrial
evolutionary connection ever existed between the plant and animal
kingdom, and within each of these kingdoms there could well also be
separate lines of 'creation'. On the other hand our theory does not permit
dispensing with terrestrial evolutionary connections between most of the
outer thinner banches of Figure 6.3. The reason is not the usual
morphology of the taxonomist. Morphological similarities could arise
between terrestrially disconnected lines simply because their programs
contained common sub-routines (all sub-routines being in any case
derived from a cosmic source). Our reason lies in embryology.

As life-forms become more complex, more care is necessary in
providing for the development of the embryo, without which of course
there could be no continuity from generation to generation. Before the

emergence of the mammals, care was exercised among animals by surrounding the fertilized female sex cell by nutrients, growing in quantity along the sequence from Figure 6.6 to 6.10. This shows, to our gastronomic advantage, in a hen's egg, and it shows overwhelmingly to the eye in an ostrich egg. Now there is no way, except by being carried in Tommy Gold's spaceship (chapter 2), that an ostrich egg from space could reach the Earth's surface in a viable form.

We emphasized in chapter 3 that no living organism, entering the terrestrial atmosphere under the most favourable orbital conditions, could retain viability if its dimensions exceeded sixty micrometres. Strictly speaking, this condition was calculated for a microorganism *travelling alone.* For an ovum travelling within a protective matrix of inorganic material, the situation at first becomes worse with increasing size. Thus for sizes of the order of one thousand micrometres (1 mm) entry into the terrestrial atmosphere is a catastrophic affair. The heat produced is so great that the entering particle is evaporated, with the resulting hot gases emitting a brief flash of light, just as we observe with so-called 'shooting stars' in the night sky.

At first sight, therefore, the idea of encasing an ovum in a protective matrix looks hopeless. As often happens, however, there is a loophole, which is to encase an ovum in a protective matrix that is too large for more than its outer skin to be evaporated. This is the case for meteorites, which land comparatively gently at ground-level after surviving passage through the atmosphere. The outer surfaces of meteorites have experienced evaporation at high temperatures of several thousand degrees, as with the smaller shooting stars, but their interior regions have remained quite cool, cool enough for volatile materials to remain undisturbed.

Although certain types of meteorite appear to have been derived from comets, and would therefore on our point of view be expected sometimes to contain living cells, we have not hitherto made use of this possibility because of its rarity. It would be irrelevant for pathogenic attacks on the Earth such as we considered in *Diseases from Space,* where in Appendix 2 we were concerned with the entry of bacteria and viruses into the atmosphere. But with hundreds of millions of years available in the present context, the rare landings of quite large cells inside meteorites cannot be ignored. New biological categories could become established in this way, perhaps even some of the outer branches of Figure 6.3. The most likely candidate is the division *Insecta,* not of course the ostrich! Insects possess such a profusion of amazingly specialized properties that one cannot help wondering if they might not have had a separate origin.

The insects are a division of the phylum *Arthropoda,* which is to say the insects belong taxonomically to a wider set of creatures having certain common properties. Spiders, scorpions, crabs, centipedes and the extinct trilobites are examples of other arthropods. All are invertebrates with a generally similar body plan, consisting of a jointed exoskeleton that allows various segments to become adapted to distinct functions, such as feeding, movement, respiration, reproduction and sensory reception.

In orthodox biology the origin of the arthropods has been a matter of controversy between those who consider them to have been derived by an explosive 'radiance' from a single ancestral stock (an annelid, i.e. segmented, worm) and those who believe them to represent various forms of more or less parallel evolution from a number of different sources. Needless to say, fossil connections capable of resolving this difference of opinion have not been found.

While the concept of an explosive 'radiance' from an annelid worm would be consistent with the discussion of the present chapter, the following chapter considers the more radical possibility that the various lines of arthropods represent similar but distinct invasions of our planet. The discussion is given explicitly for insects, partly because more information is available in their case and partly because the insects are the most spectacular representatives of the arthropods.

8
Insects from space?

It was argued in the preceding chapter that life would best be spread throughout the universe from a simple all-purpose beginning, and by subsequently adding more complex 'programs' as environmental conditions permitted. Suppose now one were to throw overboard all the previous arguments for this conservative kind of plan. Suppose one were to try spreading complex intelligent life in forms that were mainly preset. How from a logical point of view should the problem then be tackled?

Some concessions to generality would probably be essential. One would begin by making up a roster of likely environmental conditions, classifying them according to a sensible physical and chemical scheme. Then one would design the life-forms with properties related explicitly to the various categories in the classification scheme. The situation would not be all-purpose as it was before, because in this way of doing things the environment would need to be guessed beforehand, at least in a very general way, if not in considerable detail. For example, planetary environments would surely have two categories, according to whether free oxygen was present or not. Solving the problem would probably be much less difficult if oxygen were indeed available, and perhaps a restriction to this case would have to be made.

Free oxygen implies that all-purpose life has taken root already, since free oxygen is not likely to be present except through photosynthetic production by plants. In effect, then, all-purpose life would be required as a tool. Recognizing this, one might as well go on to make a virtue out of necessity, by designing the specialized forms to be out-and-out predators. Thus the specialized forms are required to be invaders that prey on every aspect of all-purpose life.

Man has become accustomed to thinking of himself as an outstanding predator, but man is not remotely to be compared with the insects in this respect. Insects exploit every land-based organism. There is no known land-based plant or animal that is not attacked in some way by them.

It is easy to see why flight would be an obvious component of a preset plan. Free oxygen implies an atmosphere, and if there is to be an atmosphere it would be natural to make use of it as a means of locomotion.

A more difficult question is how the free oxygen should be deployed. The method used by terrestrial vertebrates is to seize on the reversible oxygen-attachment property of hemoglobin, pumping the hemoglobin in a blood-supply around the body. An alternative is to have many hollow tubes threading the body along which oxygen can diffuse directly from the atmosphere, the tubes in effect being a kind of lung permeating the whole body. It is well-known that insects and some other arthropods use this second alternative, thus setting them apart from the rest of land-based terrestrial life.

It is well-known that this oxygen-diffusion system, while working excellently on a small scale, limits the sizes of insects, in the case of long thin creatures to a length of about 30 centrimetres and for more or less globular creatures like beetles to about the size of a mouse. This further limitation appears serious at first sight, because it appears to preclude the development of intelligence, but we shall see later that such a presumption on our part may be wrong.

Keeping the size small has many advantages. It removes what would otherwise be a serious uncertainty for larger creatures, the strength of gravity in the unknown environment. Insects are almost immune to the strength of terrestrial gravity. They can ride in the wind to the summits of the highest Himalayan peaks. Gravity could be a half, or twice, what it actually is, without insects being embarrassed by the change.

Keeping the size small provides insects with their most important weapon, the ability to exist in very large numbers, almost 'astronomical' numbers. It has been estimated that there are several million insect species, totalling perhaps as many as 10^{18} individuals. The greatest numbers of species are found among beetles, butterflies and moths, flies, ants, wasps and bees. Of these, beetles have an ancient history going back to about 250 million years ago. When complete specimens of beetles are found in the fossil record they are little different from present-day forms. Fossilized flies are also believed to date back to about 250 million years ago, but flies similar to living forms do not go back more than about 70 million years. This is also the measure of the ancestry of ants, wasps, butterflies and moths, while bees are found in the fossil record only back some 25 million years.

Keeping the size small permitted the adoption of an exceedingly tough and effective body plan. A typical insect body is encased within an external skeleton made up mainly of the substance chitin, which has similarities to cellulose, the polysaccharide that gives strength to the stalks of plants and to the wood of trees. Within a skeleton of chitin, the adult insect body is divided into three main parts—the head, the thorax and the abdomen. The head supports two antennae, two compound eyes and in many insects three simple eyes, and in most insects a mouth. In some, the mouth is fused into a piercing and sucking arrangement, while in others there is a biting and chewing mechanism. The head is jointed to the thorax which has three segments, each of which bears a pair of legs, with the second and third segments (taken in order from the head) typically carrying a pair of wings. The abdomen has up to 11 segments. It houses the digestive as well as the reproductive system of the insect, and is both a chemical and a physical laboratory of great subtlety.

With this apparently simple body plan, insects have invaded every land and freshwater habitat where food is available, from deserts to jungles, from ice-sheets and glacier streams to stagnant warm ponds. Some even survive in volcanic hot springs, and others (in a larval state) can live in pools of crude oil where they eat other insects that fall in. Insects are rarely found in the sea, however, a fact which makes one suspicious of the usual evolutionary story, according to which the arthropodal ancestral stock is supposed to have emerged from the sea. If it had done so, we would have expected the ocean to have been much less of an alien environment to the burgeoning insect swarm.

It is of interest to return to the embryonic problem discussed at the end of chapter 7. We saw that safe entry into the terrestrial atmosphere places limitations on the sizes of eggs. The limitations imply no restriction, however, on the basic genetic material carried by an egg, which is always trivial in scale compared with the food supply necessary to provide for the growth of the embryo. Although quite large eggs could occasionally land on the Earth safely inside meterorites, the need for such a highly special condition would weaken the force of the invasion plan at present under discussion. Is there any way around this difficulty?

A partial solution to the embryonic problem would be to design eggs that were two creatures rolled into one. The first creature would be a comparatively simple affair requiring a minimum of food for the growth of its embryo. The function of the first creature would be to collect food to provide for the growth of the much more complex second creature. The partial solution is thus to take the problem in two stages, which is just what insects actually do. Insects go through a larval (or nymph) stage which in most cases appears to have no direct connection with the adult form. Grubs become beetles and caterpillars become butterflies.

In the usual evolutionary picture it would probably be argued that a similar need occurred at emergence from the sea into land. Quite apart from the difficulty with this idea mentioned above, the evolutionary sense seems wrong. Insects that emerged late in the fossil record (e.g. butterflies) tend to show fewer points of similarity between the larval and adult stages than insects (e.g. mayflies) which appeared long before them. This is incomprehensible on the usual picture, but on the view that the insect orders which appeared later in the fossil record arrived later from space, the sense is correct. Early arrivals have to some extent freed themselves from an initial constraint, whereas later arrivals have not yet had time to do so.

Insects are bisexual, and this raises a problem as to how eggs could be fertilized. Short of a carefully controlled laboratory procedure, it is hard to see how eggs could be pre-fertilized, since after fertilization the eggs would need to be frozen immediately to prevent embryonic development from starting up. If instead of pre-fertilization we suppose sperms as well as eggs to be space-incident, there is an apparently insuperable mathematical problem. Thus the incidence of each sex would have small probability, so that the chance of fertilization would seem to go as the product of two small numbers, yielding an implausibly small result. At first sight, therefore, the incidence of a bisexual life-form from space appears to be ruled out.

Validly we can say from this argument that only life-forms which are capable of self-reproduction can be incident from space. Yeast cells reproduce by budding, and so yeast cells satisfy this condition, as do other micro-fungi, protozoa, algae and bacteria. Very remarkably, so do insects.

Parthenogenesis is a process whereby unfertilized eggs can develop by themselves. Parthenogenesis occurs not only among insects (especially those which are recent in the fossil record, like ants, bees and wasps) but has been observed in other major invertebrate groups. Mostly it is the female which self-propagates, but parthenogenesis can in some cases produce males only, or even a mixture of both sexes. Parthenogenesis solves the mathematical difficulty of the previous paragraph. (Instead of the chance of a bisexual life-form establishing itself being determined by the implausibly tiny product of two small numbers, the chance becomes of the order of the probabilities for each sex taken separately.)

Doubtless some explanation for the existence of parthenogenesis can be invented within orthodox biology, but any such explanation will surely be very weak compared to the need for parthenogenesis in the present theory, where it is essential if complex bisexual life-forms are to be incident from space. The fact that parthenogenesis exists should give sceptics pause for thought. Many insect species occasionally adopt parthenogenesis even when bisexual reproduction is available, and some species routinely alternate parthenogenetic generations with bisexual generations.

It would be possible once more in a preset plan to make a virtue out of necessity. Because insect eggs contain nutriment only for the development of the simple larval stage, leaving the larvae to provide for the adult stage, adult females can manage to lay vast numbers of eggs, instead of being restricted like birds to only a few each year. While this ability gives insects their great weapon, number, it leads to another mathematical difficulty. How among the 10^{18} insects which now inhabit the Earth are the males and females of each of the several million different species to find each other? Every physical and chemical means, except one, appears to have been brought to bear to solve this problem.

The missing method is very conspicuous, because it is precisely the method we would adopt ourselves, recognition through the acute visual resolution of details. The method is quite workable—consider how a bird in the distance is able to tell the difference between an inorganic particle and an edible scrap of food. Insects reject this method so completely, however, as to make it appear that the rejection is deliberate. Nor would it be a good idea in a preset plan to suppose that gaseous atmospheres in all cosmic environments would always be as transparent as that of the Earth. Fine particle suspensions in an atmosphere might well produce a vague diffuse lighting (like a 'white-out' on a snowfield) in which high optical resolution would be worthless. The methods actually used by insects would work just as well in diffuse conditions as in clear light. Indeed, some of the methods work even if there is no light at all.

On the first evening after travelling from temperate northern latitudes into the tropics, one is overwhelmed by the incessant noise of insects. It becomes necessary to 'tune out' this noise, otherwise (far worse than a ticking clock) it will quickly drive you mad. Much of the uproar comes from crickets, grasshoppers and cicadas, which are among the most ancient of insects. Grasshoppers saw away like violinists with wings against a series of pegs in their hind legs. Crickets rub a scraper on one wing against another wing. Cicadas have a drum in the abdomen which they can click at oscillation rates of up to 600 per second. Crickets and grasshoppers have sound receptors in their legs, while cicadas have ear drums on either side of the thorax. The purpose of the uproar is both for mating and for a general sorting and grouping of the various species. From a physical point of view, no very large quantity of information can be conveyed by such means, which is presumably why these insects keep up their racket so continuously. They keep it up so loudly because their auditory receptors are much less sensitive than the human ear, except at very high frequencies.

The hum of the mosquito has been mentioned before. The pitch of the hum is a much less broad-band affair than the sawings and clickings of crickets, grasshoppers and cicadas, and is apparently as precise as a tuning fork. Males respond to the hum emitted by females, probably through vibrations set up in their antennae, and they do so with greatest sensitivity at a pitch of about 400 cycles per second, somewhere near A in the central octave of the pianoforte.

Whereas human sensitivity to sound falls away steeply as the frequency rises above 10,000 cycles per second, insect sensitivity in many species appears to increase into the ultrasonic range. Sensitivity persists in some species well above 50,000 cycles per second. Night-flying moths respond to frequencies in the range 40,000 to 80,000 cycles per second, it has been suggested, as a protection against marauding bats. Whether such very high frequencies are used for navigation is unclear.

Insects attract each other, like-to-like by smell as well as by sound. Stories of male butterflies being attracted to a female from distances of several miles are well-known. The great sensitivity of insects to smell, however, appears to be confined to a group of substances known as the pheromones. Tested against the human on scents such as those of various kinds of orange, the honeybee has been found not to be outstandingly superior to man, except perhaps at distinguishing a particular scent among a medley of smells. With the pheromones, on the other hand, it is a different story. Whereas the bee is exceedingly sensitive to the pheromone emitted by the queen bee, humans cannot detect this substance by smell at all.

With these preliminaries out of the way, we can think about a more ambitious plan for invading a wide range of cosmic environments. Instead of requiring insects to find food simply by chance in each environment, suppose we arrange for each environment to be invaded also by plants that are specifically adapted to the needs of the insects. The problems discussed previously would then be greatly eased, and much more complex insect 'programs' could be contemplated. But a new difficulty arises, namely of establishing the plant invaders in competition with already existing plants.

It will clearly be a great advantage to the new plants to have a direct pollen vector from one to another of them, instead of needing to rely on pollen being blown in the wind—the enormously wasteful system used by the old plants. For this, the new plants can make use of the insects themselves as vectors, so establishing a logical closed loop. Once again, however, there is a problem of how to get the system started. To break into the loop it is essential to begin with the plants, for without the plants the new more complex insects cannot function. How, then, is the direct pollen vector to be established for the first of the plants needed to set up the loop? Inefficiently, but feasibly, with the aid of the already existing simpler insects, and perhaps also by mammals that have developed through all-purpose evolution. Thus the new plants will need to offer themselves to animals in general as better eating material than the old plants, a disadvantage certainly, but perhaps not a disadvantage so great that it over-compensates for the advantage of direct pollination.

This scenario agrees with the facts. The flowering plants do not appear in the fossil record until about 130 million years ago. Their tenure was little more than a toehold until about 100 million years ago, when in a geological flash they spread over the entire world. Thus the *Encyclopaedia Britannica* remarks:

> They evidently formed only small populations, and were represented by a comparatively small number of individuals (until 100 million years ago) when one of the most sudden and fundamental transformations of terrestrial plant life occurred, and in the course of a very short interval in geological time the flowering plants came to be distributed throughout the world, quickly reaching the Arctic and Antarctic regions. They appeared in great diversity and quickly became dominant . . . .

It appears no accident that one of the most remarkable and presently most widely distributed of insects, the ants, appeared about 100 million years ago, thus permitting the logical loop described above to be established.

Just as the profusion of insect species created a major recognition problem, so a recognition problem had to be solved in order that the logical loop would function efficiently. Pollen from a particular plant species had to reach other members of the same species, and not be spread around higgledy-piggledy among all plants. Thus particular plants had to attract particular insects. To do so it was necessary that an insect could distinguish a wanted plant from an unwanted one. Since visual recognition by a high resolution method had been rejected from the insect plan, signalling by colour was adopted.

By equipping both plants and insects with the same genes for colour, all manner of problems could indeed be solved. Insects could now recognize each other by colour as well as by smell and sound. By plants mocking insect-colours, and sometimes even mocking the actual appearance of an insect, the problem of insect-plant association could be solved. Moreover, by insects mocking plant-colours, camouflage protection against predators could be achieved. Butterflies and their caterpillar larvae exhibit this colour symbiosis to perfection.

In our view it is just as impossible that the necessary colour genes could have been found by Darwinian selection as that basic biochemical substances like the enzymes could have been found terrestrially. What could happen by natural selection, and what very likely did happen, was that among a number of cosmically-provided colour 'programs' for both insects and plants, those which did not suitably match each other were eliminated by extinctions. It was a case of finding by trial and error which associations fitted and which did not. It is no surprise that even to this day examples of colour-matching by natural selection can still be found. What would indeed be a surprise would be for such matchings to involve the discovery of new colour genes, rather than being variations in the expression of already existing genes.

For creatures depending on fine visual resolution such as birds and ourselves, it is not an advantage for the eyes to be sensitive to too wide a range of colour, because of the optical problem of bringing light of different colours to the same focus, the problem of chromatic aberration which plagued generations of telescope builders. But for insects dependent on colour and not on fine resolution the maximum possible colour range would be desirable. It is easy to see therefore why insect vision extends into the ultraviolet, whereas the vision of terrestrial animals produced by the all-purpose evolution of chapter 7 does not.

There would be no way to know in advance the full colour range permitted by a particular environment, and so one would sensibly choose a wider range than might turn out to be required, as indeed the range of *Drosophila* extends to ultraviolet wavelengths as short as 2537 Ångströms (as we saw at the outset in chapter 1), to wavelengths which do not exist naturally on the Earth.

Among the many more complex programs permitted by the association of insects and flowering plants, perhaps the most remarkable has been the emergence of social insects—termites, ants, wasps and bees, to name the common forms. So much has been well and clearly written on the behaviour patterns of social insects that we confine ourselves here to an amazing new discovery concerning bees.

A series of controlled experiments has shown that bees can perceive the direction of the Earth's magnetic field.[1] This remarkable perception appears to owe its origin to small quantities of aligned magnetic material that have been isolated from the abdomens of bees. Bees seem to use this perception to augment their other senses mainly for navigational purposes. In the absence of other cues they also seem, quite remarkably, to set their rhythms by the daily regular oscillations of the terrestrial magnetic field. (For instance, a bee in a black box to which an Earth-like magnetic field oscillating with a period of 6 hours is applied would tend to adopt a day-night rhythm accelerated by a factor of 4.) It is implausible to suppose that such a sharp perception of a magnetic field could have evolved in a terrestrial context. As far as we can see, there is really no need for it, and there could have been no adequate evolutionary pressures to direct its discovery. It would seem much more likely that the magnetic sense of bees is derived from a more positive need in an environment far removed from the Earth.

But we are not really concerned in this chapter with arguing the case against Darwinism—there was plenty of such discussion already in chapters 6 and 7. We are concerned here with what we regard as a more subtle issue, namely the question of whether all complex terrestrial life-forms have been derived by the cosmic all-purpose evolution discussed in chapter 7, or whether in the case of insects the evidence points rather to an explicit invasion of the Earth by an already programmed creature.

How does the situation now stand? By no means every aspect of insect life can be regarded as preset, since some terrestrial adjustment of insect programs appears certainly to have taken place. Colour matching was already an example. Another example concerns the polarization of the daylight sky. It is an advantage of the compound eyes of insects that they permit polarization to be used for navigational purposes. To maximize this advantage it would be desirable for insect eyes to be most sensitive at the wavelength where the sky happens to be brightest. When the Sun shines through clear dust-free air, the sky appears blue to our eyes, which means that the sky is brighter in blue light than it is in green, orange or red light. From this simple observation we might guess that the sky would appear still brighter if we could see it in the ultraviolet. This

[1] J.L. Gould and J.L. Kirschvink, *Science*, vol. 201, 1978, 1026.

expectation is correct. The sky is indeed brightest in the near ultraviolet, at a wavelength of about 3600 Ångströms, which is just the wavelength at which insect eyes are most sensitive. Since this value of 3600 Ångströms depends on the intrinsic colour of sunlight, it is clear that insect eyes are adapted rather precisely to the light of the Sun, and so to the particular situation as it exists in the solar system.

Against this point, however, provision for the adjustment of eye pigments to whatever light spectrum insects happened to encounter would be an obvious provision that one would expect in a plan of any subtlety. Not much can therefore be made of it.

When insects impact the windscreen of a rapidly moving car their blood is scattered on the glass surface. Common experience shows the blood (used by the insect to transport various chemicals around the body) to be of a pale, whitish colour. Because it lacks hemoglobin it is not red. So it would seem that insects normally have no use for hemoglobin. Yet hemoglobin is indeed found in the blood of larvae of certain species that inhabit the muddy bottoms of pools where oxygen is scarce. Thus the globin genes are evidently present in the insect genome, even though for the most part they go unexpressed. How much this fact can be construed as an objection to the above discussion is unclear. Certainly if a simplistic decision were taken that insects should invariably derive their oxygen by gaseous diffusion there would be no reason for the inclusion of globin genes. But if a watery medium with little dissolved oxygen in it were also contemplated, the globin genes would then be seen to be a compelling necessity.

We come now to what for us is a strong argument for the existence of an overt plan of planetary invasion. Most multicelled animals have wilted badly and in many cases hopelessly under the impact of man. Yet insects have not wilted at all. In spite of man having spent a great deal of time, effort and money in devising the most sophisticated attacks of which he is capable, the effect on the burgeoning insect population is at best one of containment. It is believed that for all our command of the physical environment, we have so far been unable to exterminate a single insect species.

Not even one species among millions! How could this possibly be? The fossil record shows insect species to be essentially unchanging over time-scales of 50 million years or more. Many insect patterns were set in the age of reptiles, before the dominance of mammals. How, then, could the emergence of so formidable a creature as man have been anticipated beforehand? By blind luck yet again, like the minute chance of finding the crucial genes on which life depends, the enzymes, the t-RNA's, the histones. . . ?

One has only to watch the brilliant aerobatics of a flock of swallows as it hunts insects in the twilight of a warm summer evening to see that a similar problem arises between birds and insects. When one considers the amazing visual acuity of birds, together with their swift mobility, it is astonishing that insects have been able to survive against them. If the fossil record did not show the evolutionary status of insects to be nearly static, if they were evolving rapidly in their forms against a selection pressure imposed by birds, the situation would still be most remarkable.

The forms of mayflies and dragonflies were set more than 100 million years *before* the emergence of birds. Yet without change they survive the biological challenge, and they do so even though from a purely visual point of view they are almost blind to the attacks of their predators.

The situation points clearly to one of two possibilities. Either we are dealing with an overt plan invented by an intelligence considerably higher than our own, an intelligence which has foreseen all our chemicals and flamethrowers, or the insects have already experienced selection pressure against intelligences of at least our level in many other environments elsewhere in the universe.

There is a curious variant of the first possibility. Could the insects themselves be the intelligence much higher than our own? We are so conditioned to thinking that the intelligence of a species can be exemplified by an individual member that it is hard to assess a situation in which each individual might show little intelligence, but in which the combined aggregate of individuals might show much. Yet it is so in our own brains, where no individual neuron can be said to display intelligence but in which the aggregate of neurons constitutes exactly what we understand by intelligence.

The static nature of insect societies goes against this thinking. If an enormous intelligence inhabits the beehives of the world, we might expect more evidence of its presence. But this may again be to endow an opponent with our own restless characteristics. Perhaps concealment is an essential tactic. Perhaps the intelligence is static because it understands the dictum of sagacious lawyers: 'When your case is going well, say nothing.'

The insect case is indeed going well. Along with the chemicals and the flamethrowers, there are nuclear bombs also. Insects are highly resistant to X-rays and other forms of ionizing radiation. Insects can frequent dumps of radioactive waste without harm. Nor are the plants on which insects feed harmed at all by radioactivity. This sets the scene for the future. From nuclear war only one creature will profit hugely, the insect. Insects may be close to inheriting the Earth without struggle. It may well seem that man arrived in a brief moment, and then disappeared even more swiftly than he came.

9
Convergence to God

At this point, if only in the interest of retaining our credentials as scientists, it might be as well to stop. After all, no biologist has cause for complaint with the present situation. Since for a century or more biologists have believed that the Earth provided an environment sufficient for the origin and evolution of life, they can hardly grouse at the scheme of Figure 4.1, which on a galactic scale gives of the order of 10^{11} times the environmental variety of the Earth, and which on a universal scale gives of the order of 10^{20} times as much.

Yet if we are to maintain a proper scientific outlook, the numbers calculated in chapter 2 have to be faced at some stage. We showed there that a random shuffling of amino acids would have as little chance as one part in $10^{40,000}$ of producing the enzymes. It is usual to attempt to side-step this difficulty by arguing that the first enzymes in the first life were much shorter in their polypeptide lengths, and so were much less improbable to come by. The idea is for the first life to evolve by natural selection, with the enzymes growing in length and becoming more complex, until eventually they reached their present forms. There is nothing in this hand-waving beyond attempting to argue that the 2,000 or so enzymes are built from a much smaller number of basic components, with each component of a simple structure. Whether or not this is so can be decided by reference to the actual amino acid sequences on the enzymes themselves. While there are some repetitions of structure (trypsin and chymotrypsin were mentioned as examples in chapter ?) we think it safe to say that if so remarkable a suggestion were true it would long since have been discovered. Besides which the same problem applies widely to other complex biosubstances such as the histones.

When one considers the need for a program to control the behaviour of cells, the problem is aggravated. Everyone who has actually set up a sophisticated program for a normal computer (not using provided languages like Basic or Fortran) will agree, we think, with our experience that the writing of sub-routines is the least part of the job. The hard part lies in the logic of the main program. In the biological case, the enzymes, histones, . . . , are only the sub-routines. The main program remains, and likely enough this is the really awkward part, a part that is probably much less likely to be discovered by random processes than the complex biosubstances on which our probability estimates have been based, much less likely than one part in $10^{40,000}$.

Given an atlas showing the amino acid sequences of all the enzymes, human biochemists could construct them with complete accuracy, thereby demonstrating the enormous superiority of intelligence allied to knowledge over blind random processes. Nor would the atlas need to be very large, not more than a single volume of the *Encyclopaedia Britannica*. Indeed, within the compass of that encyclopaedia one could specify all the structural and functional sequences of the DNA within a human cell; and, given sufficient effort, all the complex biochemicals needed by the cell could be constructed. And if the cell program were known, quite likely it too could be given a material representation. In short, it is not grossly beyond human capacity to make a functioning cell, if we had the knowledge of how to do it.

Any theory with a probability of being correct that is larger than one part in $10^{40,000}$ must be judged superior to random shuffling. The theory that life was assembled by an intelligence has, we believe, a probability vastly higher than one part in $10^{40,000}$ of being the correct explanation of the many curious facts discussed in preceding chapters. Indeed, such a theory is so obvious that one wonders why it is not widely accepted as being self-evident. The reasons are psychological rather than scientific.

Something of this kind was implied by the special creation theory— God created each individual life-form. When biologists began to see clearly that individual life-forms were not entirely separate from each other, as the original form of the special creation theory required, but that evolutionary connections existed between some of them, the impulse was to swing to the opposite extreme. The whole of the special creation theory was thought to be wrong and there was a general revulsion among scientists against it. In effect, because the details were seen to be incorrect, the fundamental idea that life was created by an intelligence was also rejected.

This forced a reliance on the inorganic processes of 'Nature'. Somehow a brew of appropriate chemicals managed to get together, the organic soup, and somehow the chemicals managed to shuffle themselves into an early primitive life-form. From then on, all appeared to be plain-sailing, natural selection operating on randomly generated mutations would do the rest.

Already in the mid-nineteenth century, however, it was seen that the chemical shuffling part of this argument was weak. Thus Charles Darwin wrote: '. . . if (and oh what a big if) we could conceive in some warm little pond, with all sorts of ammonia and phosphoric salts present, that a protein compound was chemically formed ready to undergo still more complex changes . . .'.

Although the difficulty was admitted, just how big an 'if' would be needed (one part in $10^{40,000}$) was not understood in the nineteenth century. As the enormity of the supposition was slowly revealed in the present century, there was an attempt to evade this difficulty through the invention of pseudo-science.

The pseudo-science had its origin in the second 'law' of thermodynamics. The second 'law' of thermodynamics is not a mathematical law of physics in the usual sense, not like Maxwell's equations of electromagnetism or Einstein's of gravitation which relate physical quantities at each individual point of space and time. The second 'law' is distributed over a volume of space (a closed box) and it refers to what happens in that volume in an extended interval of time. Thermodynamics is an empirical science, with the 'laws' of thermodynamics expressing certain broad features of the way complex material systems are observed to behave. One way of stating the second law is to say that a particular quantity called 'entropy' never decreases.

The problem as it seemed to many scientists was to explain in terms of the real laws of physics (like Maxwell's and Einstein's) why entropy never decreased. Perhaps the most penetrating analysis of this question was given in 1909 by the German mathematician, C. Carathéodory. His work did not commend itself generally, however, because it did not arrive at the wanted result. What Carathéodory proved was that either entropy could never decrease or it could never increase, but he couldn't say which! To agree with the observed situation it was still necessary to make an arbitrary choice.

Nowadays it is understood where the trouble lay. There is no such thing as a closed box isolated from the rest of the universe. If one thinks, for example, of a hot gas within a box, it is impossible to contain the heat energy of the gas as was assumed in 1909, because there is no such thing as perfectly insulating material from which the walls of the box can be constructed. Thus the gas inside will lose heat energy unless there is a compensating inflow from the universe outside. This means that a crucial condition assumed in Carathéodory's argument—the gas neither gains nor loses energy—is only possible if the box is suitably connected to the outside universe.

Of course people were not so innocent in 1909 that they believed in perfectly insulating material. What they argued was that the walls of the box could be made of sufficiently good insulating material so that the percolation of heat through it was small and could be ignored. This, however, was just where they went wrong. The percolation of heat may be small but it is sufficient to tip the balance between Carathéodory's two possibilities. The situation as it is understood today is that nothing internal to the box can lead to the second law of thermodynamics. It is the manner of connection of the box to the outside universe which determines the second law. With the advantage of hindsight one feels that perhaps even in 1909 this resolution of the problem should have been perceived. Since the second law does not refer to the individual points of space-time, but to a volume of space-time, it is really an almost obvious guess that the volume in question must be the whole universe. As mathematicians would say, the second law must be a global law.

Without the advantage of this perception, some scientists began to hunt around for what they felt to be a new deep principle applicable to the behaviour of material systems inside closed boxes which would decide the entropy issue from within, some deep principle that imposed itself on all material systems of sufficient complexity. The argument was of the variety which a later generation of physicists came to describe as 'pulling oneself up by one's bootstraps'. There had to be an explanation of the facts, therefore there had to be a 'principle', so no matter how great the imperfections of one's theory, it had to turn out to be correct in the long run, much the same bootstrap as is still used for Darwinism.

So it came about that the concept of a scientific unicorn which would one day turn out to be a real animal was born. The trouble with unicorns is that once you have one of them, pretty soon you have a whole herd. If a unicorn could decide the second law of thermodynamics, why not have another unicorn that could shuffle the amino acids into the enzymes? Not an intelligent unicorn like a human biochemist, mark you, but a unicorn of deep principle. Somehow matter of its own accord would shuffle itself into the enzymes, because of a deep over-riding principle in the nature of things. Except that *Nature* was the word used instead of *God,* the idea was really the same as the older religious concept it was supposed to replace.

Trotting immediately on the heels of the unicorn of deep principle was the unicorn of Darwinism. As we have seen in former chapters, there are so many flaws in Darwinism that one can wonder why it swept so completely through the scientific world, and why it is still endemic today. The reasons were several.

Natural selection as proposed by Edward Blyth in 1835 was a sound and solid idea. Only those who suffered from extreme religious prejudice could fail to accept its validity. Yet it was precisely those sufferers who chose to argue the case against Darwinism and to do so in the full glare of public controversy. Their defeat on the natural selection issue was inevitable, and when the public saw them in disordered flight it seemed only a matter of commonsense to accept the validity of the whole Darwinian story. There was no general perception that the real issue of controversy, as it had existed decades earlier between Blyth and Darwin, had still to be resolved.

The difficulty for the few who wished to come to grips with the real question of whether random mutations and natural selection had been sufficient to explain the origin of species, and by implication the origin of life, which Darwin maintained but which Blyth did not, was that in the nineteenth century the theory was impossible to quantify. Before modern microbiology, the evolutionist simply pointed to the long time-scales of geology and there was then no way to demonstrate that it would need a time-scale $10^{40,000}$ times as long to produce the effects that were being claimed.

Undoubtedly, however, the biggest thing going for Darwinism was that it finally broke the tyranny in which Christianity had held the minds of men for many centuries. Christianity as it is practised today is a rather mild social philosophy, but in medieval times it bestrode in the most dreadful way the whole range of intellectual thought. It did this by imposing on the brain a set of concepts that were false and by then insisting (on pain of extreme punishment) that all subsequent thinking be made consistent with those false concepts. A similar but less extreme situation pervades Marxist nations today. Today, a dissident can attempt to escape by crossing the 'iron curtain'. In medieval times there was no iron curtain, because there was then nowhere to escape to.

By the nineteenth century the tyranny had gone, otherwise *The Origin of Species* would have met a similar reception to that given to Copernicus' *De revolutionibus orbium caelestium libri VI*. But although Christianity had lost its thumbscrews and other instruments of torture, it still retained respectability. You could hardly be considered a gentleman unless you were a Christian, and if you wished to attend the University of Cambridge in order to study mathematics you had to pass an examination in 'Paley's Evidences' and to attend compulsory chapel. It was this less extreme position that was broken by Darwinism. It gradually became moderately respectable not to be a Christian, and it gave to those who were earnest in their thinking the confidence to rid themselves of the false concepts which had formerly been so stultifying and so mentally painful.

William of Ockham (c. 1285 to c. 1349) is often referred to nowadays as a prophet of scientific method, mostly, we suspect, by scientists who have not read anything of what he actually wrote. Ockham was a good example of how the brain of an intelligent man could be tormented into absurdities by the need to conform with Christian concepts. Typical of the issues on which his mind was occupied, in 1330-1331 Ockham preached a number of sermons in which he proposed that the souls of the saved did not enjoy the vision of God immediately after death but only after they were rejoined with the body at the Last Judgement.[1]

Needless to say, this was hardly an appropriate frame of mind in which to approach the delicate problems of scientific method.

We mention William of Ockham because doubtless there will be some who will be thinking of using his razor to slash the arguments of this book. 'Ockham's razor' would seem nowadays to mean just what you want it to mean, in the style of Humpty Dumpty. We give two examples that have been used against us in the past, and then we give our own interpretation, after which we leave it to medieval historians to decide which, if any, of the three is right.

'If there is already a theory which has not been disproved to the satisfaction of the scientific community, don't advance a new theory.' If this was what Ockham meant, then so far as science is concerned, his opinion was worthless. If Einstein had adopted such a position he would not have advanced his theory of gravitation, because Newton's theory was not considered by scientists in general to be disproved. Newton's theory was in excellent agreement with many thousands of facts. There was only one small detail which might have been considered a discrepancy, and this small detail did not seem important to most astronomers and physicists until well *after* Einstein published his theory.

---

[1] The non-political works of William of Ockham are in the course of publication by the Franciscan Institute, St Bonaventure, New York. The first volume in a projected 25 volume series appeared in 1956.

'If there are two theories, one simpler than the other, the simpler one is to be preferred.'

At first sight this does not seem quite so bad, but a little thought shows that our tendency to prefer the simpler possibility is psychological rather than scientific. It is less trouble to think that way. Experience invariably shows that the more correct a theory becomes, the more complex does it seem. Quantum electrodynamics is the only theory we have that is probably completely correct, and quantum electrodynamics is highly complex. So this second interpretation of the elusive 'razor' is also worthless.

Our own interpretation is:

'Be suspicious of a theory if more and more hypotheses are needed to support it as new facts become available, or as new considerations are brought to bear.'

This interpretation leaves Darwinism in a poor way, for this is exactly what has happened to Darwin's theory over the 120 years since it was announced in *The Origin of Species.* The disposition is to argue: 'Darwinism is correct. Therefore any hypothesis essential to its support must also be correct.' This, in our view, is what Ockham's razor enjoins one not to do.

While on this small historical diversion, it is worth adding a few remarks concerning the evolutionary theory of J.B. de M. Lamarck (1744-1829) which preceded Edward Blyth by about a quarter of a century. Lamarck argued that characteristics acquired by the parents tended to be passed to their offspring. If you spent your life in a gymnasium your children were more likely to be athletic, or if you spent your life in intellectual studies your children were more likely to be born geniuses. It would easily be possible to produce impressive statistics in support of this idea, and in an age innocent of the lies which can be told by statistics, people could just as easily be deceived into believing it to be true. The point of course is that individuals who are inherently weak physically do not choose to spend their lives in a gymnasium. Nor do the less intelligent wish to spend their lives in intellectual studies. In a complex society with many diverse occupations available, most people select the work they can do best. In effect, the social environment itself acts as a mechanism of choice. It is no wonder, then, that the abilities of the offspring turn out to be generally correlated with the activities of their parents.

The statistical trap in Larmarck's idea was not too difficult to spot, a trap that was avoided by Darwinism. It is likely therefore that, following the publication of *The Origin of Species,* little more would have been heard of Lamarck if an ironical social development had not occurred in the latter half of the nineteenth century. Just as Nature is said to abhor a vacuum, so it seems that societies abhor fluid intellectual situations. People seem more contented in their lives if there are simple broad issues to which they can devote themselves with great passion. With the breaking of the intellectual bonds imposed by Christianity there was a mighty casting around among those with a lot of time on their hands for some new form of intellectual bondage. Quite a lot of solutions were tried. Some tried throwing aside the trappings of civilization, returning to the simple life along with the animals in the field, but the attempt to crop grass failed because we do not possess the right kind of bacteria in our digestive system. The winner proved to be Marxism in some societies, and its more pallid relative, socialism, in others.

The early Marxist idealism was very different from its later rigid reality, just as early Christianity had been very different from the later rigid reality of the medieval world. To begin with, all the talk was about the dignity and fulfilment of man. To anyone imbued with such ideas, the teachings of biology must have seemed appalling. Yet Darwinism could not be denigrated, since it was Darwinism that had destroyed the hated rival, Christianity. In this dilemma a loophole was perceived, Lamarck-ism.

It was eminently in keeping with the socialistic ideal for the earnest efforts of the parents to be rewarded by the improved abilities of their offspring. How much better that Beethoven should have been born to parents who had worked as earnest and humble musicians (especially the humble part) than that he should be born to a tubercular mother and a drunken father. (That the genetics of biology permit a great genius to arrive out of an unfavourable social situation is actually a superb inspiration, comparable to the inspiration of the late quartets of Beethoven himself.)

The idea was really the same as the Christianity it replaced, of course, except that instead of being rewarded in 'heaven' for one's efforts in life, it was now to be one's children that were to be rewarded. Indeed, there was little that was different in the new socialism from the old Christianity, except the removal of the concept of *God,* which ironically was the sole concept of any importance in the old religion.

The loophole was to combine Darwin and Lamarck. Natural selection was seen to be fine, provided it operated on acquired characteristics in the sense of Lamarck. Many attempts were made to prove that acquired characteristics could be inherited. Experiments to this end continued well into the second half of the twentieth century.

Although the issues were sometimes subtle, the claimed proofs of Lamarckism have always been shown false, to a point where not even the leaders of the Soviet Union dare attempt any longer to maintain a Lamarckian position, although doubtless they will return to it should they succeed eventually in dominating thought throughout the world.

The failure of Lamarckism to prove itself has been taken, not just as a disproof of Lamarckism, but as a proof of Darwinism. If the two had been shown to exhaust the possibilities, the logic of this inversion would be justified. Of course no such mutual exclusivity has been shown. We are not faced by an either/or situation, and so the inversion is just as false as were the claimed proofs of Lamarckism. Nevertheless, the downfall of Lamarckism certainly played an historical role in convincing biologists that Darwinism must be correct.

To what has been said in our attempt to understand why Darwinism has achieved such a strong grip on scientific thinking we must add the great force of educational continuity. It is hard in later life to doubt those basic tenets of intellectual thought which all one's teachers have accepted without question. We have had quite a number of encounters with biologists who have remarked much as follows: 'I will admit that your views are consistent with the facts, and that they even have a certain driving logical quality which normal theory lacks, but I just cannot bring myself to face the upheaval in my thinking that would follow if I agreed with anything you say.'

Once the whole of humanity becomes committed to a particular set of concepts, educational continuity makes it exceedingly hard to change the pattern. You either believe the concepts or you will inevitably be branded as a heretic. The Protestant reformers of northern Europe broke free from control by the Catholic Church of Rome after a long and bitter struggle, but they did not do so by breaking free from Christianity itself. The issues of great controversy were trivial, such as how many angels can cluster on the head of a pin, or like the inconsequential issue raised in the sermon of William of Ockham. Real escape for the individual was impossible.

Escape is comparatively easy, however, when humanity is divided over a controversial issue, because if you are thinking of changing your mind on it there is always someone in the opposite camp with whom you can talk about your problems. This is why controversial situations are healthy, why a non-controversial society is unhealthy, and why the leaders of the Soviet Union are so concerned that 'dissidence' be suppressed throughout the territories controlled by the KGB and the Red Army. This is the reality towards which Marxism has led. Past history and educational continuity are undoubtedly two reasons for the hold which Darwinism maintains on the scientific world, but we would be doing our biological colleagues serious injustice if we maintained they were the only reasons. Undoubtedly there is also a clear appreciation of what the alternative to remaining within the safe smokescreen of Darwinism would mean.

It is as if one were sitting in the mist on a little pile of snow. The mist clears and instead of finding ourselves on some friendly eminence as we had hoped, we are suddenly on the North Wall of the Eiger. Once we say that only through intelligence can the enzymes and other biochemicals be assembled, we find ourselves with a long and desperate climb ahead, a climb to safety which it may take many generations to achieve.

The first few steps are not so bad, however, just as the first few steps on the Wall of the Eiger might not be so bad. We happened to mention Copernicus' book *De revolutionibus* in the above discussion. In his book Copernicus abandoned unequivocally for the first time the notion that the Earth is the geometrical centre of the universe. The usual biological theory, a terrestrial organic soup plus a miraculous assembly of complex biochemicals leading to the origin of life, is an attempt to maintain an essentially pre-Copernican position, with the Earth as the biological centre of the universe. It will be no bad thing to abandon this position. The avidity with which some people seek to maintain it is clearly analogous to the passionate attempts of the sixteenth century to retain the Ptolemaic theory of the pre-Copernican era. If thereby man's horizons turn out to widen as much as they have done over the three centuries which separated *De revolutionibus* from Hubble's discovery of the expansion of the universe, nobody should be unduly surprised.

The next step is to face up to a chicken-and-egg problem. If life existed already, it would not be hard to imagine an intelligence somewhere in our galaxy (or elsewhere in the universe) deciding to assemble the enzymes. But what would be the point of it if the enzymes existed already in the functioning of the life-form itself? Even if we could show that the assembly was for some purpose, we would be no further forward in a logical sense, we would still have to explain how the enzymes came to be assembled in the first place in order to permit our postulated life-form to function.

Here we might seek to play a naive trick. We could say that the assembly of the enzymes had then become a problem for the other life-form, not for ourselves. Stated this way, the trick seems absurd, but it is really just this same trick that is played in all religions, namely to displace all problems to *God* and then refuse to discuss them any further.

To be consistent logically, we have to say that the intelligence which assembled the enzymes did not itself contain them. This is tantamount to arguing that carbonaceous life was invented by a non-carbonaceous intelligence, which by no means need be *God*, however.

Silicon is an atom similar to carbon, and there has often been speculation on whether a form of life based on silicon instead of carbon might exist. Yet if one attempts to follow a similar chemical system for silicon—silicon sugars, silicon nucleic acids, silicon proteins and so forth—the idea soon grinds to a halt, because it is easy to show that, although generally similar, silicon is less versatile chemically than carbon. It is therefore hardly conceivable that a siliceous form of life could have preceded the carbonaceous form, at any rate if the two are thought of in terms of similar chemistries.

But what if we forget about chemistry and think of electronics? Then it is silicon that wins handsomely over carbon. Everybody has heard nowadays of the silicon chip, but nobody considers the possibility of a carbon chip, for the thing would be an electronic absurdity. So what if our progenitor were an extremely complex silicon chip?

One thing looks right about this idea. It would not be possible for an intelligence, however great, to generate carbonaceous life without performing an immense amount of calculation. While the blueprints for all the enzymes, and quite a number of other crucial biochemicals, too, could be specified within the compass of a single volume of the *Encyclopaedia Britannica,* distinguishing that particular volume by calculation from all the $10^{40,000}$ volumes with incorrect specifications would be a task far beyond human capability. The best way we know to perform the necessary calculations would be through the silicon chip. For doing, for getting about, for being active, carbonaceous life is the best, but for swift calculation, and perhaps for thinking too, a siliceous form of life could be greatly our superior.

If we could estimate the amount of calculation in a quantitative way, the idea would be testable in principle, because then we could arrive at a specification of the computational power that would be needed. Unfortunately, however, not enough is yet known of the mathematical detail of complex chemical structures to decide whether or not the calculations would be feasible, even for the silicon chip. We can only speculate that they might be.

The crucial point of the argument is that intelligence and swift calculation could face up to an analysis of the potentialities of a wide range of possibilities, whereas mere random trials with actual chemicals could never do so. Much the same is true even on the level of the human chemical industry, which would soon descend to chaos if it elected to depend only on a random throwing together of chemical substances.

From the point of view of the thinking silicon chip, what would be the point of it? Tools. Tools for the control of large-scale astronomical processes, as one of us has suggested elsewhere.[2] And a very powerful tool, in the sense that we humans have now given rise in our turn to the silicon chip itself. So one would have the sequence (in which the arrows mean 'leading to'):

silicon chip → carbonaceous life → silicon chip

by which the silicon chip would succeed in spreading itself.

So much is conceivable within our present knowledge. Yet if this were all, there would be a lack of grandeur about the idea that would surely come as an anticlimax to the story we have been attempting to develop. To proceed further we can either write

? → silicon chip → carbonaceous life → silicon chip

where ? is an unspecified intellect, or we could take ? as our immediate progenitor: ? → carbonaceous life → silicon chip

losing some definition of the problem but adding to our sense of dignity.

The interest now focuses on ?, so that it becomes somewhat irrelevant which of the above sequences we choose. Remarkably enough, we have evidence of the existence of ?. Life depends on oxygen and carbon being roughly equal in their cosmic abundances. If either one dominated markedly over the other, life would not be possible. The requirement is for oxygen to be rather more abundant than carbon, which is exactly how things actually stand.

Both of these elements are produced from helium by nuclear reactions that occur inside stars, the details of which are quite well understood. So far from the abundances coming out correctly in an unavoidable way, it turns out that getting conditions right depends on a couple of curious properties, one a property of the carbon nucleus, the other of the oxygen nucleus. (Both the 7.65 Mev level of the carbon nucleus, and the 7.12 Mev level of the oxygen nucleus, must be tuned very closely indeed to these particular energy values.) If we did not know from laboratory experience that everything is the way it should be, if in ignorance we had to set a chance of things coming out appropriately, the chance might be estimated at about 1 part in 1000.

In the past the favourable nuclear properties had been thought of as curiosities, lucky accidents of physics, without which life could not exist. It was as if a child had twiddled the tuning knob of a radio receiver, and

[2] F. Hoyle, *The Relation of Biology to Astronomy*, University of Cardiff Press, 1980.

then when you yourself switched on the receiver the tuning just happened by accident to be on exactly the station you were seeking—except that the accident was there twice, once for the carbon nucleus and once for the oxygen nucleus.

Over the quarter of a century since these properties of oxygen and carbon were discovered, the disposition of astronomers has been to shy away from the thought that the situation might be deliberate. There is a mental trick one can easily play, both here and in respect of several other favourable provisions of physics. The trick begins with a correct statement, namely that if things had been otherwise there would have been no life, in which case we ourselves would not have been around to think about the problem. So far so good. Then comes an inversion of the logic. Because we *are* here, the argument continues, the favourable provisions of physics *must* hold, and therefore no problem exists. Our existence fixes the physics. The argument is really not much better than the logic of the following question and answer:
Q. Why is A taller than B?
A. Because A is 6 ft 1 in. tall whereas B is only 5 ft 9 in.

If a random inorganic origin of life could have been maintained, then one could not have made too much out of this situation. Everything would be the result of chance. Once we see, however, that the probability of life originating at random is so utterly minuscule as to make the random concept absurd, it becomes sensible to think that the favourable properties of physics on which life depends are in every respect deliberate.

The measure of intelligence needed to control the properties of the oxygen and carbon nuclei would be exceedingly high. The so-called coupling constants of physics are numbers which appear in science empirically. That is to say, they are numbers that we determine by reference to observation rather than from logical argument. The basic unit of electrical charge (the charge of the electron) is one such number. So far as the consistency of physics is concerned the unit of electrical charge could apparently have an infinity of values other than the value we assign to it from observation.

It is the coupling constants which determine the favourable aspects of physics, like the favourable properties of the carbon and oxygen nuclei. It would be through exercising control over the coupling constants that an intelligence might determine a wide range of features of the universe. The remarkable chemical behaviour of the carbon atom and the remarkable electronic properties of the silicon chip are other crucial examples of properties which might be controlled in this way.

It will be apparent that we have moved on now to an altogether higher level of intelligence than the silicon chip. Calculating the properties of the enzymes would surely be an amazing achievement, judged from the human level, but likely enough it would seem rather a simple matter to an intelligence that could control the coupling constants of physics. In astronomical terms, control over the origin of life is probably equivalent to controlling processes on the scale of stars, whereas control of the coupling constants is probably equivalent to controlling processes on the scale of galaxies.

We can now suggest the equivalence:

Control over the coupling constants of physics ≡ ?

The disposition next is to add ? ≡ *God,* and so to be done with it. But such a slapdash conclusion leads to difficulties. It leaves the relation of *God* to the universe still quite unclear. As we contemplate the resulting situation, ? would still be only a part of the universe, *within* the universe as one might say, and so *God* would only be a part of the universe, surely an unsatisfactory conclusion.

We must also guard against an emotional difficulty of the human mind. Most people are happy with a situation in which there is just one intellectual level higher than the human, leading to the satisfactory relationship *God* → man. This relationship, analogous to parent and child, is emotionally pleasing but hard to defend logically. From the point of view of the dog there would be a three-tiered structure, *God* → man → dog, and from the point of view of the meanest sparrow that falls there would be a considerable many-tiered structure.

People in the past did not think in quite this biological sense, but they could see at a glance that society was many-tiered in its relationships. Instead of just king → subject, there was king → nobles → knights → squires → yeomen → folk.

The conflict between the emotional requirement for a two-tiered structure and the logical requirement for a many-tiered structure led to wild confusions in the religious systems of the past. The Greeks were never able to get their religion under control. The system (in which the arrows indicate progression from higher beings to lesser beings, here and in the following sequences)

shadow figures → Zeus → main gods and goddesses
e.g. Chronos

→ subsidiary gods and goddesses → kings → people
caused them a lot of trouble, but not remotely to be compared with the

problems of the Catholic Christian, who is required to master a sequence that runs approximately as follows:

God → Christ → archangels → angels → saints
→ the beatified → pope → cardinals → priests → people.

Although these particular sequences are plainly unacceptable, the sequence idea is far better logically than the over-simple two-tier *God* → man concept. To avoid the ludicrous, we must write:

. . . → ????? → ???? → ??? → ?? → ? → man → . . .

The dilemma of religion is that the sequence is meaningless unless we attach explicit significance to at least one of its members (to the left of man) and yet attempts to do so have always led in the past to absurdities.

The position of most scientists can, we think, be said to accord with one or other of the following three points of view:

(1) there is no such sequence;
(2) the correct sequence is the simplistic one, *God* → man;
(3) there is such a sequence, but since we know nothing about it there is no point in discussing it.

Our opinion is that all of these are wrong. The correct position we think is: there is such a sequence, and among the question marks to the left of man there is a term in the sequence, an intelligence, which designed the biochemicals and gave rise to the origin of carbonaceous life. Still further to the left there is another still higher level of intelligence that controlled the coupling constants of physics.

This may seem a grey form of religion, not at all suited to the wearing of gaudy clothes or to parades in the streets on saints' days, but it is far better to be in with a chance of being modestly right, instead of being faced by the absolute certainty of being overwhelmingly wrong.

Where does the sequence going to the left stop? It doesn't. It goes on and on and on, with ever-rising levels denoted by more and more question marks. But like a convergent mathematical sequence of functions it has an idealized limit, with the property that by going far enough to the left the terms differ by as little as one pleases from the idealized limit. It is this idealized limit that is *God,* and *God* is the universe:

God ≡ universe

The logical system is now closed, leaving no inherently baffling questions, except one. The remaining conundrum is:

Why is there anything at all?

Of course one can play the usual anthropic trick of saying that if there was nothing, there would be nobody to ask the question. But this is an evasion, not an answer. Nor would it be an answer for a physicist to argue that the universe is created by particle pairs emerging from the vacuum, because the physical properties of the vacuum would still be needed, and this would be something.

The old religions were at their best when they kept to generalities such as 'man was created in the image of God' (to give this particular idea its Judaic form, although it probably dated from much earlier times). The sequence displayed above would not make sense unless each level of intelligence was contained in all those levels to the left of it in the sequence. It is therefore almost inevitable that our own measure of intelligence must reflect in a valid way the higher intelligences to our left, even to the extreme idealized limit of *God*. Our defects lie in a restriction of the intellect, not in a failure to reason correctly within our own lights. Otherwise our situation would be hopeless, and indeed so would be the situation of every member of the sequence.

The calculation of the properties of the enzymes is an example of a restriction of the intellect. We ourselves could not cope with the enormity of detail involved in this particular problem, but we can easily conceive of the calculation being done, and we can even understand the general way in which it would go, and the physical principles on which it would be based. Modern developments in physics may not be too far from the day when we can understand how the coupling constant problem might be fixed. The situation therefore appears to be that, while we cannot actually emulate what the intelligences far to the left in the sequence are doing, we can understand in a non-detailed way what kinds of monkey business they are up to. And this is really what science, properly understood, is all about.

There are those who think that the way to establish a connection with a higher intelligence is to go riding around the galaxy in spaceships, or to listen for coded radio signals emanating from other planetary systems in the galaxy. In the past we have had some sympathy with the second of these notions, but have always rejected the first as woefully slow, crude and profitless. Now we are inclined to think the second is also unnecessarily cumbersome. If our ideas are correct, there must surely be a multiplicity of clues around us here on the Earth's surface, clues to the identities of the intelligences immediately to the left of us in the sequence, probably as far as the intellect which calculated the properties of the enzymes. It would be more sensible, for example, to broadcast clues on the unexpressed DNA of yeast cells than to go through the ponderous technology of radio transmissions (discussed in our former book *Lifecloud*). In a search for such clues it would of course be necessary to acquire the DNA before it became garbled by too many copying errors.

If we invert the religious dictum that man was created in the image of *God,* to the extent of attributing to all higher intelligences the chief characteristics of the mentality of man, then it seems possible that higher intelligences have a pronounced sense of humour. The clues might then have been distributed in the fashion of the child's game of 'hunt the thimble'. Attempting to cast ourselves in the role of a higher intelligence, it would surely suit our sense of humour to lock up clues in the genome of the California redwood. Or if one were more inclined to farce than wit, one might lock up clues in the social behaviour of insects. Either possibility would be much simpler than travelling around the galaxy.

There is in many people a strong emotional conviction that some higher intelligence must be around us all the time. The idea is far better than its attempted realization as the occupants of unidentified flying objects, for example. Such vague gropings will not do, of course, but the idea may still be true in the sense that a multiplicity of clues are lying around, waiting for us in the hunt-the-thimble game. A still better idea, again in essence an old religious idea, is that the important connections along the sequence

$$\ldots \rightarrow ???? \rightarrow ??? \rightarrow ?? \rightarrow ? \rightarrow man \rightarrow \ldots$$

occur in thought. Not by any means all thoughts, although the whole process of consciousness probably has a profound cosmic significance. The connections of the sequence are more likely to be restricted to those sudden flashes of perception that have made so much difference to all the main trends of human thought, the conversion of Paul on the road to Damascus.

Conclusion

This book has presented the relation of biology to astronomy on a wider perspective than did our previous writings. Here we attribute both the origin of life and its continuing evolution to cosmic influences.

Although our point of view is anti-Darwinian and is in a sense a return to the concept of special creation, it is not the old concept of special creation. If we define 'creation' to mean arrival at the Earth from outside, the unit of creation in our picture is the gene, not the working assembly of genes that we call a species. Which assemblies of genes survive and which do not is decided by the environment of the Earth. The potential of life is cosmic but its realization is terrestrial.

We have received hints and even warnings from friends and colleagues that our views on these matters are generally repugnant to the scientific world. We in our turn have been disturbed to discover how little attention is generally paid to fact and how much to myths and prejudice.

It is not hard to find writings in which the myth is stated that the Darwinian theory of evolution is well proven by the fossil record. But one finds that the higher the technical quality of the writing the weaker the claims that are made. The imperfections (discussed in chapter 6) are blamed in even the best texts, however, on the incompleteness of the fossil record. Yet if one persists by consulting the geological literature the truth eventually emerges. The fossil record is highly imperfect from a Darwinian point of view, not because of the inadequacies of geologists, but because the slow evolutionary connections required by the theory did not happen. Although paleontologists have recognized this truth for a century or more, they have not been able, in spite of their status as the acknowledged experts in the field, to make much of an impression on consensus opinion.

The rate at which genes coding for polypeptides (e.g. hemoglobin) undergo mutations has been measured in the laboratory to be not greater than 10^{-5} per generation, which implies a rate for changing individual amino acids in the corresponding polypeptide chain of as little as 10^{-7} per generation. If only ten amino acids of particular kinds are necessary at particular locations in a polypeptide chain for its proper functioning, the required arrangement (starting from an initially different arrangment) cannot be found by mutations, except as an outrageous fluke. Darwinian evolution is most unlikely to get even one polypeptide right, let alone the thousands on which living cells depend for their survival. This situation is well-known to geneticists and yet nobody seems prepared to blow the whistle decisively on the theory. If Darwinism were not considered socially desirable, and even essential to the peace of mind of the body politic, it would of course be otherwise. If our own theory contained such a howler many voices would be raised in chorus against it.

We did not arrive all in a moment at the position described in this book. We had no thought in the beginning that the small track we were then following (organic materials in interstellar space) would broaden eventually to become a major highway. Only gradually with the discovery and dovetailing of many facts did the overall picture at last become evident. But in that dawn of certainty, in what might have been a moment of satisfaction, we hit a difficulty that knocked the stuffing out of us. No matter how large the environment one considers, life cannot have had a random beginning. Troops of monkeys thundering away at random on typewriters could not produce the works of Shakespeare, for the practical reason that the whole observable universe is not large enough to contain the necessary monkey hordes, the necessary typewriters, and certainly the waste paper baskets required for the deposition of wrong attempts. The same is true for living material.

As our ideas developed, a monstrous spectre kept beckoning. Just as the brain of Shakespeare was necessary to produce the famous plays, so prior information was necessary to produce a living cell. But information from where? From some pre-existing life-form, one is tempted to answer, but this would be to incur the ire of Tommy Gold, who tells the following story.

A male lecturer had spoken about the nature of the Earth and planets. Afterwards, an old lady came up to him from the audience, claiming she had a theory superior to the one he had described. 'We don't live on a ball revolving around the Sun,' she said, 'we live on a crust of earth on the back of a giant turtle.'

Wishing to humour the old lady the lecturer asked, 'And what does this turtle stand on?'

'On the back of a second, still larger turtle', was the confident answer.

'But what holds up the second turtle?' the lecturer persisted, now in a slightly exasperated tone.

'It's no use, mister,' the old woman replied, 'it's turtles all the way down.'

Which epitomizes the conundrum of life. So long as living cells come from pre-existing cells we are calling on support from another turtle. The issue is where do the turtles stop? The conventional answer is that the turtle pile floats on a sea of organic soup, an answer as scientifically improbable as Tommy Gold's story makes it sound.

In his denial of the doctrine of spontaneous generation, Louis Pasteur set himself on the side of the little old lady, as did Lord Kelvin in the remark quoted above (chapter 3):

> . . . we all confidently believe that there are at present, and have been from time immemorial, many worlds of life besides our own. . . .

So too did the German scientist Hermann von Helmholtz (1821-1894), who stated the issue in 1874:

> It appears to me to be a fully correct scientific procedure, if all our attempts fail to cause the production of organisms from non-living matter, to raise the question whether life has ever arisen, whether it is not just as old as matter itself, and whether seeds have not been carried from one planet to another. . . .

The old lady's point of view forces one to the strictest form of steady-state cosmology, in which there are no beginnings, only unbroken chains stretching infinitely far back into the past. While the steady-state cosmology is now in much better fettle than its detractors over the past fifteen years would have one believe, we hesitate to adopt such a strict interpretation of the theory. Certainly over a time-span much greater than that of big-bang cosmology, much greater than 10 billion years, the universe may well be approximately unchanging. We doubt, however, that an unvarying steadiness can be carried into an infinite past, as Kelvin and Helmholtz in effect supposed. Since we regard the universe as evolving, but on a time-scale much longer than 10 billion years, we are still left with the problem of what ultimately supported the old lady's pile of turtles.

Conclusion

From the beginning of this book we have emphasized the enormous information content of even the simplest living systems. The information cannot in our view be generated by what are often called 'natural' processes, as for instance through meteorological and chemical processes occurring at the surface of a lifeless planet. As well as a suitable physical and chemical environment, a large initial store of information was also needed. We have argued that the requisite information came from an 'intelligence', the beckoning spectre.

To be sure, the books in a library contain information. Yet we do not think of a book as an 'intelligence'. A further quality is needed to define intelligence, namely the ability to act on information which a book alone cannot do. As well as providing information, our spectre is required also to act on it, which is why we refer to the thing as an 'intelligence'. It may be objected that a computer can act on information. Precisely so, which is why in chapter 9 we discussed the possibility that the unseen face of our ghost might be a computer console.

Such a pedestrian identification would not end the matter, however, for it would transpose the problem of the origin of life to the problem of the origin of computers. We simply exchange a pile of turtles for a pile of elephants. The last part of chapter 9, which we shall not repeat here, gave our attempt to grapple with this transposition, and with the problem of infinity, the ultimate compass of the universe.

Appendix I

A HISTORICAL NOTE ON THEORIES OF EVOLUTION

Does it really matter whether our ideas about the past are right or wrong? In our view there is no absolute answer to this question. Historical truth as one perceives it on any issue must of necessity be relative to one's concern with that issue. If our concern happens to be nil, it hardly matters if our ideas are wrong or indeed if we have any ideas at all. For the authors, the details of the negotiations which led up to the Diet of Worms are irrelevant, and it does not embarrass us that we know nothing about them. For the medieval historian, however, instructing a class of students avid to learn about the Diet of Worms, it would be otherwise. Likewise it is more relevant for scientists to be correct in matters of scientific history than it is for the non-scientist.

Disrespect for accuracy in matters of scientific history is common in the current practice of science. A happy-go-lucky attitude to history is only too likely to go hand in hand with a happy-go-lucky attitude to facts that are of relevance at the frontiers of science, while deliberate connivance at historical inaccuracy is only too likely to be first cousin of the suppression and distortion of facts, a disreputable practice which unfortunately is not without practitioners.

In popular opinion, scientific ideas about biological evolution changed decisively with the publication of Darwin's *Origin of Species* in 1859. This would be the opinion too of most workers in the physical sciences. Perhaps for laymen and physical scientists it does not matter too much that this belief is at variance with the truth. It does matter, however, that most students of biology are taught the same incorrect history.

On the occasion of the centennial jubilee of the *Origin of Species*, a well-referenced account of developments in evolutionary biology before 1859 was given by L.C. Eiseley (*Proc. Am. Phil. Soc.,* 1959, vol. 103, 94-158). The situation can be summarized in the following quotation:

> As one looks back over the curious and intertwined history of Darwin and his associates one is struck more and more by the smooth unbroken evolution of the natural selection concept from the time of the eighteenth-century writers onward. An enormous body of myth has obscured this process. In this hundredth anniversary year since the publication of the *Origin*, the number of eulogies, addresses, and similar encomiums is burying ever deeper the true story of the past.

The first reference to 'natural selection' traced by Eiseley is given in Patrick Matthews's book on *Naval Timber and Arboriculture* published before Darwin set sail in 1831 on the famous voyage of the *Beagle*. Matthews writes of 'this natural process of selection'. Some four years after returning from the voyage, Darwin set out his ideas in an essay in 1842, changing the words slightly to 'natural means of selection'. The same form appeared again in a second essay written in 1844, but is then shortened to 'natural selection' as the writing proceeds. Neither essay was published in Darwin's lifetime. Both appeared in 1909, however, under the title of *Foundations of the Origin of Species*. Many years later, after the publication of the *Origin*, Darwin complained at the widespread failure of the public to understand the meaning of 'natural selection'. If he had kept to Matthew's clearer words, there would have been less room for public misunderstanding.

Edward Blyth was born in London on 23 December 1810. His father died in 1820, leaving a widow and four children in straitened circumstances. Not surprisingly, Blyth's education proceeded erratically, especially as he seems to have had little interest in studies other than natural history. In such matters, however, he was both a dedicated observer in the field and a voracious reader, especially in the British Museum.

Of special relevance to this note are two of Blyth's many papers. The papers in question appeared in 1835 and 1837, both in *The Magazine of Natural History*, a leading journal of the day. The references are vol. 3 (1935), pp.40-53, and vol. 1 (1937), pp. 1-9; pp. 77-85; pp. 131-141, the second paper appearing in three parts. Eiseley reproduces them verbatim, together with a paper of 1836, and of their significance he remarks:

> As one leafs through Blyth's small papers . . . one is amazed by the ideas which reappear in [Darwin's] trial essays of 1842 and 1844 and which Darwin never altered throughout his life.

The similarities had not escaped two earlier commentators: H.D. Geldart, 'Notes on the life and writings of Edward Blyth', *Trans. Norfolk and Norwich Naturalists Soc.,* vol. 3, 38-46, 1879; and H.M. Vickers, 'An apparently hitherto unnoticed anticipation of the theory of natural selection', *Nature,* vol. 85, 510-511, 1911. Until Eiseley's publication in 1959, however, biologists had not thought fit to explore the implications of these references.

Blyth never enjoyed good health, and in 1841 he was given medical advice to emigrate to a warmer climate. A small post was obtained for him as Curator of the Museum of the Royal Asiatic Society of Bengal, where he remained until 1862. Thus Blyth, poor financially and in weak health, was far away in India during the critical twenty-two years which separated his 1837 paper from the publication of the *Origin of Species* in 1859. In 1860, Darwin wrote in a letter to Charles Lyell:

> I have had a letter from poor Blyth of Calcutta, who is much disappointed at hearing Lord Canning will not grant any money . . . . Blyth says (and he is in many respects a very good judge) that his ideas on species are quite revolutionized. . . .

The money Lord Canning would not grant was Blyth's pension. In Blyth's obituary note (*Journal of the Asiatic Society of Bengal,* August 1875) it is said that eventually a pension of £150 a year was conceded, due to efforts in London on his behalf by a Dr Falconer and Sir P. Cautley. Blyth died on 27 December 1873.

His paper of 1835 describes conservative natural selection, the process whereby a species clearly adapted to an environment does not lose that adaptation. In the paper of 1837, Blyth puts the crucial question:

> A variety of important considerations here crowd upon the mind; foremost of which is the enquiry, that, as man, by removing species from their appropriate haunts, superinduces changes on their physical constitution and adaptations, to what extent may not the same take place in wild nature, so that, in a few generations, distinctive characters may be acquired, such as are recognised as indicative of specific diversity? It is a positive fact, for example, that the nestling plumage of larks, hatched in a red gravelly locality, is of a paler and more rufous tint than in those bred upon a dark soil. May not, then, a large proportion of what are considered species have descended from a common parentage?

So already in 1835-6, while Darwin is still away on the *Beagle,* the crucial question has been asked. Blyth has now only to plough ahead regardless, arguing that the known diversity of species forces an affirmative answer to the question, and the *Origin of Species* is his for the taking. Instead he draws back, answering no, for the following reason. He considers the case of a species well-adapted to a particular environment which exists only over a limited geographical area. Blyth now argues as follows. If indeed the species could modify itself by natural selection, it would surely change gradually as the environment changed at the boundary of the area, and by this means adaptation and number would be preserved. Instead, his observations show the species obstinately maintains itself, with the consequence that adaptation is lost and the population declines across the boundary of the area.

Edward Blyth was not alone in having posed the crucial question. The great geologist, Charles Lyell, had come to the same point, and he too had drawn back, for what was perhaps an even more powerful reason than Blyth. Lyell believed deeply in the uniformity principle of James Hutton, according to which the present terrestrial environment is representative of the environment in past geolgocial ages. So for Lyell, natural selection would need to operate against an essentially constant environment. How, then, could well-adapted species ever change drastically, as it would be necessary for them to have done if they had evolved from a common ancestral stock?

Think of a flat plain out of which rise a number of mountains, with each distinct species, past and present, represented by its own particular mountain. Think of height above the plain as a measure of adaptation. Natural selection operates to keep each species on the very summit of its own mountain. Consider now the possibility of some variant of a species going from one mountain to another, as would be necessary if gross evolutionary changes of one species into another were to occur. The journey could be made in gradual steps only if the variant in question started by going downhill. That is to say, the variant would need to start by becoming disadapted, in which case it would immediately be wiped out by competition from the non-variant form still occupying the original mountain summit, and which therefore would be better adapted to the environment.

It was into this situation that Darwin returned from the voyage of the *Beagle*. It is not only inconceivable in principle that Blyth's papers, published in a major journal, would have escaped Darwin's notice, but as Eiseley points out at length there is ample evidence from Darwin's essays of 1842 and 1844 that he had studied Blyth's work closely. So the crucial question lay there, directly on his desk.

It was a normal impulse for Darwin to seek an affirmative answer to the question, opposite to the answers given by Lyell and Blyth, since by doing so he had an original line of thought to pursue. The story is that Darwin set himself on this path because of experiences on the voyage of the *Beagle*, but the popular story is to be doubted. By 1838, Darwin had not yet formulated his position in any detail. Indeed, the essays written in 1842 and 1844 were surely the first attempts to explore the situation in depth.

Yet the *Beagle* was important from the beginning in a psychological respect, for the facts which Darwin had acquired on the voyage showed Blyth and Lyell to be wrong on one issue, and if Blyth and Lyell could be wrong there they could be wrong on the key question. Both of Darwin's 'opponents', as he came to think of them (and as he wrote later without specifying names), had assumed that a species well-adapted to a particular environment would be able to cope successfully with competition from outside that environment. While this might seem a common-sense presumption there was no strict logical necessity for it. If A and B are two equally matched football teams, and if they play together on B's ground, the usual presumption is that B will win. On a general statistical basis the presumption is correct, but it is not so in every individual case. And if A is a strong team from a large city, with B only a small-town side, the presumption becomes that A will win in spite of B having the ground advantage.

This was just what Darwin had found. The extinction of the marsupial animals of S. America was the oustanding example. From perhaps seventy million years ago until about twenty million years ago S. America was separated from the other continents of the world. At the time of the separation the placental mammals had not yet emerged and the earlier marsupials were dominant. Over the intervening fifty million years the placental mammals emerged, however, in competition with marsupials in the old world and N. America but not in S. America. Thus when a land bridge between S. and N. America was formed some twenty million years ago there was a second conflict between the two kinds of fauna, this second conflict being fought on ground to which the marsupials had become specifically adapted. Yet the placental mammals won.

Darwin also observed that distinct varieties on offshore islands tended to be replaced by forms from the mainland, an example of A, a large-city team, being opposed to a small-town team B, with the match played on B's ground. Such matches occur in cup-ties, and occasionally B happens to win. Even so, small-town teams never appear in the cup-final.

It is indicative of Darwin's character that he does not seem to have told Lyell or Blyth of their mistake. Thus Lyell in a letter many years later to Alfred Russel Wallace:

> When I first wrote, thirty-five years ago, I attached great importance to preoccupancy, and fancied that a body of indigenous plants already fitted for every available station would prevent an invader, especially from a quite foreign province, from having a chance of making good his settlement in a new country. But Darwin and Hooker contend that continental species which have been improved by a keen and wide competition are more frequently victorious over an insulated or more limited flora and fauna.

This was in 1867. Lyell is sufficiently surprised by his mistake to be writing about it eight years after the publication of the *Origin of Species*, which he would scarcely be doing if he had been told of it in 1838. Darwin by his own account was a voracious reader of other men's work, obviously of Blyth's papers and certainly of Lyell's *Principles of Geology*. It was not in his character, however, to make a return for what he received, a trait that is fortunately unusual in science, but which we have ourselves encountered in modern times. From the beginning, Darwin seems to have cast Lyell and Blyth in the role of 'opponents', although nothing of which we are aware points to the opposition being reciprocated.

So in 1838 Darwin is poised to strike out on a new line, but he finds himself pinned back by the arguments of Blyth and Lyell. He has no answer to them, and exclaims:

> If one species does change into another it must be *per saltum . . .*
> [*by a jump*].

As we have shown in this book, here was the correct answer. Species change in jumps, not in small steps. They go from one mountain summit to another by way of a 'helicopter' provided by a supply of cosmic genes. This was the complete answer to Lyell's difficulty, and it agreed with Blyth's requirement that species be clearly demarcated. Yet it was a route which Darwin would not take, for it was essentially the 'special creation' position adopted by Lyell and Blyth themselves. No laurels could be won that way.

So it was essential to keep to evolution in small steps, and in doing so a devastatingly wrong path was taken. Eiseley's otherwise excellent essay is unfortunately marred by his adherence to the conventional belief that Darwin's route was correct. Eiseley details the steps that were taken over the succeeding years. The first is the one discussed already above. The second is:

(2) Oscillations of sea-level on islands and continents create shifting conditions—isolation, faunal migrations, etc. This creates opportunities for *new* selection of *new* characters, not just the *conserving* selection of Blyth.

It it easy to see why Darwin could not publish this argument in 1842 or 1844, for it would have stood little chance against the opposition of Lyell, who would undoubtedly have argued (correctly) that, while such sudden changes admittedly do arise from time to time, they are peripheral to the total world situation. Indeed, this second argument largely contradicts the first, according to which it is the selection of species in the widest continental environment that ultimately dominates the world situation.

Eiseley's third summarizing point is:

(3) New conditions, Darwin maintained, increase the tendency for mutations to appear. This happens under wild conditions in a form analogous to what Blyth contended was the case among domestic animals. Similar migrations of faunas and floras into new environments would promote mutation.

Because he believes in the ultimate correctness of Darwinism, Eiseley forbears to remark that this point is Lamarckian.

Eiseley as well as Darwin is in error over the fourth point:

(4) Darwin argued cogently that *no* environment is completely static and therefore renewed selection is going on even under superficially uniform conditions.

This would only be true if the breeding group were infinitely large. The breeding groups that Darwin used as examples, especially among animals, are actually rather small, 10^4 to 10^5 perhaps. For such groups small changes in the environment have no selective effect, because the situation is dominated by stochastic processes within the breeding group. Returning to our analogy of well-adapted species represented by the summits of mountains rising above a flat plain, if the summit of a particular mountain is dome-shaped, stochastic variations can take the *whole* of the species occupying that mountain some way downhill. The mathematical situation is that the smaller the breeding group the steeper the slope the species can descend. This behaviour, which was not properly understood until even less than twenty years ago (and which was perhaps seen most clearly by Japanese biologists), can form the basis of an attempted answer to Lyell. For a sufficiently small breeding group, and for a mountain whose slopes are not too steep, a complete descent of a whole species to the surrounding plain is possible. The species can then 'random walk' around the plain until it finds a new mountain up which it can ascend, by which time its form has become drastically changed.

This was the basis of the argument of Dr Ohno, discussed in chapter 7. The trouble, as we saw there, is that the chance of finding a new mountain by random walk alone is much too small.

The fifth point is:

> (5) Darwin invokes a law of succession to explain why the living organisms of a continental area ordinarily show the closest affinities to the extinct life of the same region. This anatomical relationship between the past and present is only explainable on the basis of evolution.

There is confusion of logic here. The facts suggest that species change their forms markedly with geological time, but there is nothing to show they do so by small steps. Indeed, the fossil record points clearly to the changes being *per saltum,* not in small steps, as we showed in chapters 6 and 7.

> (6) Finally, in the *Origin of Species,* Darwin devotes an entire section to the formidable point raised by Blyth as to the rarity or absence of transitional varieties between the various modern faunas. It was Blyth's argument, based on modern observation, that transitional varieties could not have subsisted. Referring to his 'opponents' with deliberate vagueness in Chapter 6 of *The Origin,* Darwin confesses [that] 'this difficulty (Blyth's) for a long time quite confounded me.' Nevertheless, he came to believe the difficulty could be explained.
>
> We tend, Darwin argued, to think too much in terms of recent climate and geographical gradation and then expect life to grade itself as imperceptibly. Actually life is historical. It may intrude into new areas. Species compete and sharply limit each other's expansion. Intermediate varieties, therefore, tend frequently to contract their range and disappear. Thus Darwin, after long study and analysis, broke through the seemingly durable species-barrier which Blyth had, paradoxically, created upon the basis of natural selection.

This concludes Eiseley's summary of the arguments that separate Darwin in 1859 from Lyell and Blyth in 1838. We have preferred to quote Eiseley, a Darwinian himself, rather than make our own summary, to avoid an accusation of presenting a one-sided analysis. So what in the end did it amount to? Points (1) and (5) are irrelevant to the main issue. Points (3) and (4) are wrong. Point (2) is peripheral, and as for Point (6), the second paragraph sums up the whole situation for what it was—hand-waving.

The factors that separated 1859 from 1838 were not scientific. They were that Lyell had grown old, Blyth was a forgotten man in India, and the world in general had come to want to believe in evolution. The year 1859 was nicely poised, between an earlier age in which the majority of educated people would instinctively have opposed evolution by natural selection to an age in which people, after an initial crisis of thought, were prepared to accept it. The technical issues were irrelevant to the change of outlook, which was a product of the social and economic development of industrial society. Hence no substantive arguments were required. Vague generalities, such as those of the second paragraph of point (6), were quite sufficient. And this is the way it has remained ever since.

Eiseley, although a supporter of Darwinism, devotes a great deal of his essay to a discussion of why Darwin makes no reference in *The Origin of Species* to Blyth's critical papers of 1835 and 1837, although he references Blyth in other much less important respects. The evidence does not permit of any conclusion except that the omissions were deliberate. Eiseley also wonders why Blyth did not protest about the omissions, but on this there seems no mystery to us, since it was in 1860 that Blyth was desperately trying to secure his future. To have created a rumpus over priority with a man of Darwin's position would not have seemed the best way to obtain the pension of £150 per year.

There is no law which compels a scientist to reference his sources. It is only protests from colleagues, and the fear of being treated similarly, which keep the record straight. The social situation referred to above ensured that *The Origin of Species* would not be overlooked or forgotten. On this score Darwin had nothing to fear.

The failure of biologists to insist on this matter being set right is somewhat surprising, especially as an attempt to plagiarize the work of Mendel at the beginning of the twentieth century did in fact set off a major scandal. Aside, however, from the gentle remonstrances of Geldart in 1879 and Vickers in 1911, there was nothing in a hundred years up to Eiseley's courageous essay of 1959. It would seem to us that a serious sin of omission remains to be redeemed by the world of professional biology.

We have shown in this book that the situation of 1859 was not an improvement on that of 1838. Blyth and Lyell were nearer to the truth than Darwin. When Darwin presented a paper (with Alfred Russel Wallace) to the Linnean Society in 1858, a Professor Haughton of Dublin remarked 'all that was new was false, and what was true was old'. This, we think, will be the final verdict on the matter, the epitaph on Darwinism.

Appendix 2

A TECHNICAL NOTE ON INTERSTELLAR GRAINS

In this note we present a first-order solution to the longstanding problem of the nature of the interstellar grains. The immediate reason for our confidence in this solution is the agreement shown in Figure 1 between the observed interstellar extinction of starlight plotted as a function of inverse wavelength and the calculated effects of the particle distribution discussed below. The observations represented by the points of Figure A.1 were taken from three sources and the calculations shown by the continuous curve were done using accurate Mie formulae. Both observations and calculations were normalized to an extinction of 1.8 magnitudes at $\lambda^{-1} = 1.8\mu^{-1}$, which is the extinction produced by the grains when starlight traverses a path length of 1 kiloparsec through the interstellar medium in the solar neighbourhood of the galaxy.

There is little variability in the observations at visual wavelengths taken from star to star in the solar neighbourhood of the galaxy, but there is considerable variation in the ultraviolet. The observed points for the ultraviolet are averages for a number of stars. Even so, the US and the European results still show appreciable differences, as can be seen from the distinctly displayed points in the figure.

Although the calculations which have hitherto been published have not agreed nearly as well with the observations as the curve of Figure A.1, because the definition of our particle distribution contains certainly three and perhaps four parameters, it is not possible from Figure A.1 alone to invert the logic, and to say the agreement proves the model. For this, a further strong argument is needed. Such an argument, based on the amount of material available for grain formation and on a quantum mechanical lower limit calculated by E.M. Purcell, will next be discussed.

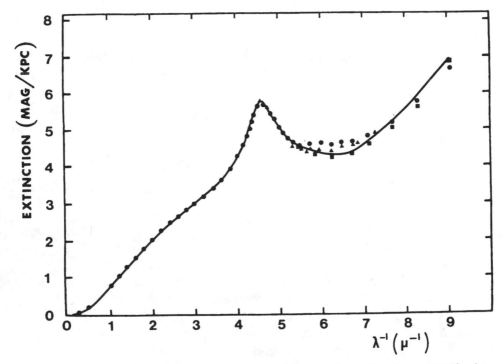

Figure A.1 Wavelength dependence of interstellar extinction normalized to 1.8 mag/kpc at $\lambda^{-1} = 1.8\mu m^{-1}$. Points are astronomical observations; solid curve is for the grain model proposed here (● Average extinction data compiled from many sources by Sapar and Kuusik,[1] ▲ ESA data from Jamar et al,[3] ■ OAO II data from Bless and Savage[2])

An upper limit to the spatial density of grains

Pure hydrogen has considerable attractions as a material for grain formation, but laboratory measurements of the vapour pressure of solid hydrogen at very low temperatures unfortunately made it necessary to rule out this possibility.[4] Neither is helium a possible grain-forming material. We are required therefore to form grains from the higher elements, C, N, O, Mg, Al, Si, Fe, to name the more important ones.

[1] A. Sapar and I. Kuusik, *Publ. Tartu Astrophysical Observatory,* 46, 1978, 71.
[2] R.C. Bless and B.D. Savage, *Astrophys. J.,* 171, 1972, 293.
[3] C. Jamar, D. Macau-Hercot, A. Monfils, G.I. Thompson, L. Houziaux and R. Wilson, *Ultraviolet Bright-Star Spectrophotometric Catalogue,* European Space Agency, Paris, 1976.
[4] N.C. Wickramasinghe and K. Nandy, *Nature,* 219, 1968, 1347.

Although hydrogen may be in combination with some of these elements, hydrogen cannot thereby add much to the mass of the available grain-forming material.

According to the well-known relative abundance table compiled by A.G.W. Cameron,[5] which we take to apply to interstellar material, the amounts of the important higher elements expressed as percentages of the amount of hydrogen are:

C = 0.45,   N = 0.16,   O = 1.08,

Mg = 0.080,   Al = 0.072,   Si = 0.088,   Fe = 0.146.

Although nearly all the C and N, and essentially all the Mg, Al, Fe, are known to be depleted from the atomic interstellar gas, and must therefore be in molecular form or in grains, about two-thirds of the O is atomic.[6,7] Thus the maximum mass density available for grain formation expressed as a percentage of the hydrogen mass density is about 1.3. The average hydrogen mass density is usually taken to be $2.0 \times 10^{-24} \, \text{g cm}^{-3}$, corresponding to a number density of 1.2 atoms cm^{-3}. Hence the maximum spatial density of available grain-forming material is $\sim 2.6 \times 10^{-26} \, \text{g cm}^{-3}$. To change this upper limit for the grain density, it would be necessary to alter either the average hydrogen number density or the relative abundances of Cameron.

A lower limit to the spatial density of grains

Using the observed interstellar extinction, together with the Kramers-Krönig dispersion relations, Purcell calculated a lower limit for the spatial density of grains in the solar neighbourhood.[8] The method assumes maximum quantum mechanical efficiency at all wavelengths and so yields the smallest grain density consistent with the physical laws. No explicit grain model is required to establish the Purcell lower limit, except that spherical grains of uniform specific gravity, σ say, were used. (Strictly, the few assumptions that have gone into this calculation[8] are only very weakly dependent on certain model parameters.) In a more recent review article, Purcell and Aannestad[9] obtained $8.6 \times 10^{-27} \, \sigma$ gram cm^{-3} for a visual extinction normalized to 2 magnitudes per kiloparsec, which is close to the normalization value of 1.8 magnitudes per kiloparsec used in Figure A.1. A slight adjustment for this small difference changes the above formula to $7.7 \times 10^{-27} \, \sigma$ gram cm^{-3}.

[5] A.G.W. Cameron, *Sp. Sc. Rev.*, 15, 1970, 121.
[6] G.B. Field, *Astrophys. J.*, 187, 1974, 453.
[7] K.S. de Boer, *Wisconsin Astrophysics Preprint*, no. 107, 1980.
[8] E.M. Purcell, *Astrophys. J.*, 158, 1969, 433.
[9] P.A. Aannestad and E.M. Purcell, *Ann. Rev. Astro. Astrophys.*, 11, 1973, 309.

Since the elements C, N, ... , form an assortment of grains with differing specific gravities, it is necessary to interpret σ as an average, weighted with respect to the abundances of the various kinds of grain. We shall see below that a third to a quarter of the carbon must be in the form of graphite, which has a specific gravity of 2.25. If Mg, Al, Si are in the form of magnesium aluminum silicates, as is usually supposed, the silicate grains would have specific gravity ~ 3.3. If Fe is metallic, as is also usually supposed, the iron would contribute particles of specific gravity 7.86.

These high values of the specific gravity imply poor efficiency in producing extinction. At first sight one might think to improve the efficiency by supposing O to be combined with H_2 to form water-ice ($\sigma \simeq 1$), leaving the remaining two-thirds of the C also to combine with hydrogen into a low specific gravity hydrocarbon. This tactic is strictly forbidden by observation, however, because interstellar water-ice would readily be detected in the spectra of all stars where extinction is appreciable, on account of its exceedingly strong absorption band at $\sim 3.1 \mu$. Since there has been no such detection, except perhaps marginally in one or two stars for which the extinction is unusually large, not more than a very small fraction of the O can be combined with H_2 into H_2O.

The natural association of the oxygen is with the ~ 70 per cent of the carbon that is not graphite. In competition for oxygen, carbon takes precedence over the other common elements, because of the very strong binding energy of the CO molecule. The first step towards forming solid material from CO is a combination with H_2, H_2CO, and then a number of units of H_2CO form either into polyformaldehyde or into sugars and polysaccharides.[10, 11] These substances have specific gravities which average about 1.5, and if we take them together with the particles discussed in the previous paragraph the average specific gravity rises to about 2. Inserting $\sigma = 2$ in the result of Purcell and Aannestad, one obtains a quantum mechanical lower limit of $\sim 1.6 \times 10^{-26}$ g cm^{-3}. This is the least grain density that can explain the observed interstellar extinction of Figure A.1.

Combining this lower limit derived from quantum mechanics with the upper limit obtained previously, we have the following inequalities that must be satisifed by any proposed model for the interstellar grains:

$$\sim 1.6 \times 10^{-26} \text{ g cm}^{-3} < \text{model density} < \sim 2.6 \times 10^{-26} \text{ g cm}^{-3}$$

We shall see below that our model requires a grain density of

[10] N.C. Wickramasinghe, *Nature*, 252, 1974, 462.
[11] F. Hoyle and N.C. Wickramasinghe, *Nature*, 268, 1977, 610.

$\sim 1.8 \times 10^{-27}$ g cm^{-3}, not much above the quantum mechnical lower limit and comfortably within the amount of material that is available. The solution leading to Figure A.1 is therefore highly efficient in the extinction which it produces. We consider it unlikely that other radically different solutions, based on solids built from C, N, O, Mg, Al, Si, Fe in the correct relative abundances, can be found. Any such alternative solution would have to satisfy the above inequalities as well as give excellent agreement with the extinction data.

Graphite spheres

Figure A.2, calculated for graphite spheres with radii 0.02 μm, shows the factor Q_{ext} by which the extinction cross-section exceeds the geometrical cross-section. The contribution of scattering, Q_{sca}, to the extinction is seen to be about 20 percent, with ~ 80 per cent of the extinction coming from absorption. The measured scattering of the interstellar grains at λ^{-1} = 4.6 μ^{-1} is actually about 35 per cent,[12] but since the other two forms of grain discussed below also contribute significantly to the scattering at this wavelength, the graphite cannot give an appreciably larger Q_{sca} than is shown in Figure A.2. This precludes graphite spheres with radii appreciably larger than 0.02 μm, while smaller spheres are excluded by the circumstance that the wavelength of maximum extinction moves shortward for smaller grains, to about 2130 Å, and the precise agreement of the graphite maximum with the observational maximum at ~ 2180 Å is then lost. Nor can the graphite grains be other than spheres, otherwise there would again be a displacement of the wavelength of maximum extinction from $\lambda = 2180$ Å. The grains must be spheres and they must have radii rather precisely determined at 0.02 μm.

In arriving at the calculated curve of Figure A.1, the density of graphite spheres was taken to be 1.79×10^{-27} g cm^{-3}, about one-tenth of the total grain density.

Small dielectric spheres

Graphite alone gives too deep a minimum at $\lambda^{-1} \approx 6.5 \ \mu^{-1}$ and too little extinction at $\lambda^{-1} \approx 8.5\mu^{-1}$ for there to be any possibility of graphite spheres being the sole cause of the extinction at these shorter ultraviolet wavelengths. A further component to the grains is needed in the ultraviolet, a need which no model can avoid. Since, moreover, there is a marked increase of the observed scattering of starlight shortward of the

---

[12] C.F. Lillie and A.N. Witt, *Astrophys. J.*, 208, 1976, 64.

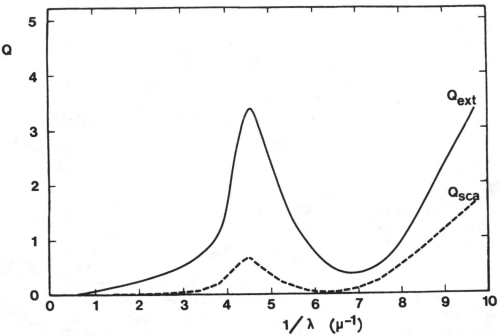

Fig. A.2   Total extinction and scattering efficiency factors for graphite spheres of radius 0.02 μm calculated using Mie formulae and the measured values of the optical constants for graphite

graphite peak, the further component must be essentially dielectric, which is to say the complex refractive index of the grains, n-ik, cannot have more than a moderate imaginary part (k not larger than ~ 0.05).

Taking the grains to be spheres, the most efficient radii for producing extinction at the shorter ultraviolet wavelengths lie in the range 0.03 to 0.04 μm. In our calculation we took radii all equal to 0.04 μm, n-ik = 1.5-0.0i, grain specific gravity 1.6, and an interstellar density of 4.29 x 10^{-27} g cm^{-3}.

In a more refined second-order calculation it would be possible to investigate the interesting region of the spectrum around $\lambda^{-1} = 6\ \mu^{-1}$ in more detail, by choosing an appropriate value of k. Using k=0, as we did in our calculation, is almost surely an oversimplification, particularly as organic materials—which we suspect the grains to be—tend to have absorptions at just these wavelengths.

If organic materials based on C, O, H are indeed involved, then a degradation of a fraction of the dielectric spheres contingent on the removal of O and H (such as occurs in the production of coal) would lead to graphitic spheres with radii ~ 0.02 μm. Thus instead of the graphite grains being an independent component of the model, there would then be a straightforward connection between them and the dielectric spheres.

Hollow cylinders

There is no possibility of explaining the visual extinction ($\lambda^{-1} < \sim 3.5\,\mu^{-1}$) with either of the two kinds of grains discussed so far—both give far too little visual extinction. Whether we like it or not, the facts show that a third type of grain must be present in interstellar space. For maximum efficiency in the visual, this third type must have a characteristic dimension of about $2/3\,\mu$m, very much larger than the first two kinds. If the third kind are spheres, then $2/3\,\mu$m is the sphere diameter, and if rods or cylinders then $2/3\,\mu$m is the cylinder diameter—there is no efficiency condition on the cylinder length.

An apparently small detail at $\lambda^{-1} \simeq 2.3\,\mu^{-1}$ in Figure A.1 has set a hitherto intractable problem. A small reduction in the upward slope of the observational points can be seen to occur at this wavelength, a reduction that Mie calculations for earlier models have persistently failed to explain. The effect is undoubtedly real and is found for stars scattered widely in the galaxy. We realized long ago that this problem might be solved if the real part n of the refractive index of the particles responsible for the visual extinction was less than 1.2, but no sensible solid material based on C, N, O, ... satisfies this condition. Recently, however, we came to believe the visual grains might be dried-out bacteria, which is to say hollow cylinders. The free space within the particles then has the effect of reducing the value of n averaged for the volume of the whole particle to less than 1.2. Using this idea we were able at last to solve the slope problem.[13]

The present calculation for the third type of particle is the same as that described in reference 13, except that we have added a small refinement, by taking the averaged dielectric constant to be 1.16 - 0.015i instead of 1.16 - 0.0i. In attacking the Purcell limit discussed above, we wished to take advantage of the available iron, and the simplest way to do so was to distribute the iron inside the hollow grains, when the Mie formulae automatically took it into account.

The particle properties were otherwise the same as we used formerly, particularly the size distribution for bacteria that was specified in reference 13 by a histogram. The average specific gravity of the hollow cylinders was 0.52, and their interstellar mass density was

[13] F. Hoyle and N.C. Wickramasinghe, *Astrophys. Sp. Sc.*, 66, 1979, 77.

7.76×10^{-27} g cm^{-3}. These numerical values refer to the dielectric material of the particles.

Adding the latter density value to those of the graphite and small dielectric spheres gives a combined mass density of 1.38×10^{-26} g cm^{-3}, and adding the further contribution of the iron and the other metals increases the density to a total of $\sim 1.8 \times 10^{-26}$ g cm^{-3}, the value stated previously.

The available density of grain-forming material was calculated above to be $\sim 2.6 \times 10^{-26}$ g cm^{-3}. If the Cameron abundances and the interstellar hydrogen density on which this estimate was based are firmly accepted, there is an unused excess of available material amounting to $\sim 8.10^{-27}$ g cm^{-3}. A fraction of this excess is still required, however, to provide for the CO gas and for other gaseous interstellar molecules. We regard the remaining residue as being present in the form of grains with sizes too large for them to have been relevant in the calculations leading to the curve of Figure A.1, as for instance eukaryotic cells with sizes of ~ 10 μm or more.

Second-order effects

We have omitted some second-order effects but have included others. One omission has been noted already, absorption in the ultraviolet by organic molecules. One inclusion was the change of dielectric constant for the hollow cylinders, from n-ik = 1.16 - 0.0i to n-ik = 1.16 - 0.015i. This latter refinement changes the visual extinction from being wholly a scattering effect to ~ 75 per cent scattering and ~ 25 per cent absorption, a change in agreement with observation.[12]

Another second-order effect, the polarization of visual starlight, has also been considered. Polarization is caused by a systematic alignment of a fraction of the cylindrical particles, the alignment being with respect to the configuration of the galactic magnetic field along the line of sight from the Earth to the star in question. The alignment is believed to be caused by a hysteresis effect of domains of iron within the particles. For reasons unconnected with this note, it appears to us improbable that all the iron is distributed uniformly among the hollow particles. We therefore assigned only 50 per cent uniformly and homogeneously among the main set of hollow cylinder particles, the ones with n-ik = 1.16 - 0.015i, and we reserved the remaining 50 per cent of the Fe for a special subset comprising only a few per cent of the particles. If the excess of iron atoms are gathered into clumps of sizes ~ 200 Å these grains could become effectively aligned.[14] This procedure gives an explanation of why

[14] R.V. Jones and L. Spitzer, *Astrophys. J.*, 147, 1967, 943.

the polarization of visual starlight usually amounts to only a per cent or two, arising from a modest fraction of particles that are well-aligned, the subset with the excess iron in the form of tiny clumps, which can be associated with iron-containing bacteria.

The excess iron in the aligned subset necessarily raises n above the value 1.16 appropriate to the main set of particles. This detail shows itself in a subtle way, in the behaviour of the polarization with respect to wavelength, which is correct for $n \simeq 1.3$, but not quite so for $n = 1.16$.

The infrared

There is no possibility at all (solid hydrogen being excluded) of explaining the extinction of Figure A.1 unless a considerable fraction of the interstellar C, N, O is condensed into solid grains. This almost forces the further conclusion that most of the grain material must be organic, particularly as water-ice grains are excluded for the reasons stated earlier. Organic material inevitably contains C-H linkages, which have an infrared absorption band at 3.4 ± 0.1 μm, the precise wavelength of the band centre being variable from one organic substance to another.

The concept of organic solids widespread in space has not lacked for critics, who have maintained the concept to be wrong because otherwise the ~ 3.4 μm C-H band would have been widely observed in reddened stars and in other non-stellar infrared sources. However, the mass absorption coefficient at the centre of the C-H band, taken with respect to the residual absorption in the distant wings of the band, is merely ~ 300 cm^2 g^{-1}, a tiny value compared to $\sim 33,000$ cm^2 g^{-1} for the nearby water-ice band at ~ 3.1 μ.[15] It needs only a few per cent of water-ice to be present in organic material for the C-H band to become relegated to a minor shoulder in the absorption spectrum of the material, and such shoulders are in fact observed in many astronomical infrared sources.[16]

For the 3.4 μ C-H band to be observable as a clear-cut absorption dip, it is essential both for water-ice to be absent and for the amount of the visual extinction to be very large, much larger even than for highly reddened stars like VI Cyg No. 12. These conditions are satisfied for the grains along essentially the whole of the ~ 10 kiloparsec line of sight to the galactic centre. The triangular points of Figure A.3 show recent observations by D.T. Wickramasinghe and D.A. Allen for the source IRS7 at the galactic centre, for which the visual extinction is thought to be about 34 magnitudes.[17] The absorption centred at 3.39 ± 0.02 μm is

[15] J.E. Bertie, H.J. Labbé and E. Whalley, *J. Chem. Phys.*, 50, 1969, 4501.
[16] K.M. Merrill, R.W. Russell and B.T. Soifer, *Astrophys. J.*, 207, 1976, 763.
[17] D.T. Wickramasinghe and D.A. Allen, *Nature*, 287, 1980, 518.

quite clear-cut in this case.

The larger circular points of Figure A.3 give earlier observations of B.T. Soifer, R.W. Russell and K.M. Merrill.[18] No absorption is seen at ∼ 3.1 μm, showing that water-ice is essentially completely absent from the whole pathlength to the galactic centre. (Water may well be present, however, in much more than trace quantities within the compact molecular clouds.)

The absorption of Figure A.3 at 3.39 μm establishes that the C-H band must necessarily be a common feature of the material of the interstellar grains, but it does not in itself demonstrate that the organic material is built from units of H_2CO, as we argued above. For this, we must consider the observational points close to 2.95 μm in Figure A.3. Whereas a pure hydrocarbon has no appreciable absorption feature there, solids built from H_2CO have such a feature derived from OH

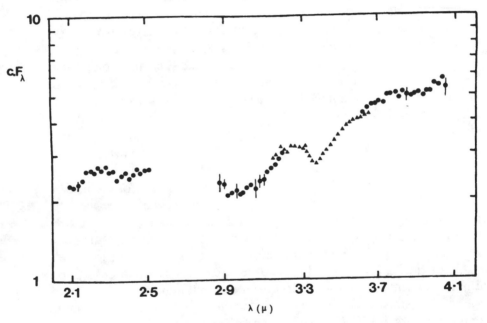

Fig. A.3   Flux in arbitrary units of the galactic centre infrared source IRS7 obtained by D.T. Wickramasinghe and D.A. Allen[17] (▲) combined with data over more extended waveband obtained by Soifer *et al*[18] for their 17″ aperture observations (●)

linkages. While observations close to 2.9 μm have to be interpreted cautiously, owing to absorptions shortward of 2.9 μm in the terrestrial

[18] B.T. Soifer, R.W. Russell and K.M. Merrill, *Astrophys. J.*, 207, 1976, 183.

atmosphere, the observations do seem to show a rather definite minimum near 2.95 μm. Moreover, one can argue with fair certainty that there must be a rise in the IRS7 spectrum shortward of 2.9 μm, simply to reach the higher flux values at 2.5 μm.

Figure A.4 shows the absorption spectrum of dry amylose (starch) obtained from laboratory data[19] by the method described by us elsewhere.[20] Although other polysaccharides are known to have absorp-

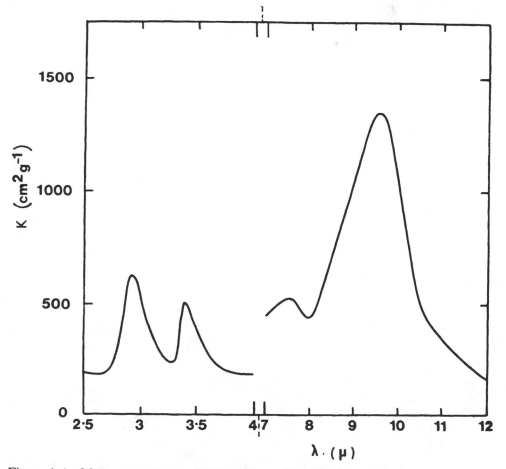

Figure A.4 Mass absorption coefficient of dry polysaccharide, using data for amylose given by Hoyle *et al.*,[19] and the procedure for estimating infrared properties for anhydrous polysaccharides given elsewhere (Hoyle and Wick-ramasinghe[20])

[19] F. Hoyle, A.H. Olavesen and N.C. Wickramasinghe, *Nature,* 271, 1978, 229.
[20] F. Hoyle and N.C. Wickramasinghe, *Astrophys. Sp. Sc.*, 72, 1980, 247.

tion bands at slightly different wavelengths, the relative strengths of the bands at ~ 2.95 μm, ~ 3.4 μm and ~ 9.7 μm are essentially as in Figure 4, which may be taken to be typical of a solid polymer based on H_2CO.

The absorption feature at ~ 9.7 μm is about five times stronger than the one at ~ 3.4 μm, both features being estimated with respect to the nearby continuum. (The absorption at 3.4 μm is taken relative to the continuum at 3.55 μm, the absorption at 9.7 μm is relative to the continuum at 8 μm. The astronomical observations are similarly differenced.) Hence interstellar organic polymers based on H_2CO would be expected to produce an absorption dip at 9.7 μm in the spectrum of a straight line infrared source that was about five times the dip at 3.4 μm. The observed dip at 3.4 μm for IRS7 is 0.22 magnitudes. Hence we expect IRS7, and other sources close to the galactic centre, to show a

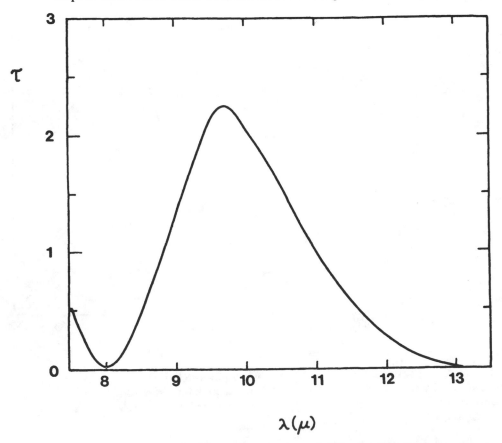

Fig A.5 The average absorption in magnitudes for the 8-12 μm band, calculated from the data for IRS 1, 2, 3, 4, 6, 8, and 10 given by Becklin *et al.*,[21] by taking the continuum to be given by a straight line joining the observed points at $\lambda = 13\mu$m and $\lambda = 8\mu$m

dip of about 1.1 magnitudes at 9.7 μm. Figure A.5 is an average of the 8-13 μm opacity for a number of such sources.[21] The average absorption at 9.7 μm is seen to be about 2.2 magnitudes, twice the expected value, showing decisively that some other agent among the grains contributes an absorption in the 9.7 μm region that is closely similar to the absorption produced by the main distribution of the grains.

We agree with popular opinion that this other agent is based on silicon, but the agent can have nothing to do with so-called 'silicates', which require each silicon atom to have acquired two oxygen atoms. It is impossible in such a picture to ensure that every scrap of SiO_2 is intimately bound with other solids, with MgO as enstatite, for example. Some free silica grains would necessarily be present, and silica has two fantastically strong infrared bands at 8.7μm and 12.7μm with absorption coefficients of $\sim$ 30,000 cm^2 g^{-1}. Since neither of these absorptions has been observed, even as tiny features in any of a large number of sources, it is clear that silicon forms SiO, not SiO_2, just as carbon forms CO, not CO_2. And just as we think polymers based on H_2CO constitute the main interstellar grain component, so we think that polymers based on H_2SiO constitute an analogous secondary component that happens to contribute equally with the carbonaceous material in the 9.7 μm region, but not in shorter-wave regions of the spectrum.

Laboratory data give 9.7 μm band strengths of 4- to 8000 cm^2 g^{-1} for polysiloxanes,[22] compared to $\sim$ 1300 cm^2 g^{-1} for polysaccharides, a ratio of about 5:1. On the other hand, the silicon-to-carbon mass ratio in the interstellar medium is about 1:5, so that the band-strength factor just happens to compensate the mass-ratio factor, which is why the $(H_2CO)_n$ and $(H_2SiO)_n$ solids give nearly equal contributions to the absorption in the 9.7 μm region.

[21] E.E. Becklin, K. Matthews, G. Neugebauer and S.P. Willner, *Astrophys. J.*, 219, 1978, 121.
[22] *Spectra and Chromatograms of Elemento-organic Compounds*, 'Khimiya', Moscow, 1976.

Index

Printed in the United States
By Bookmasters